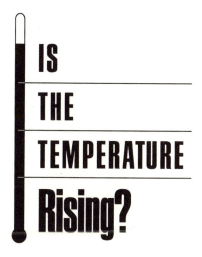

IS THE TEMPERATURE

Rising?

IS THE TEMPERATURE Rising?

THE UNCERTAIN SCIENCE OF GLOBAL WARMING

S. George Philander

PRINCETON UNIVERSITY PRESS

PRINCETON, NEW JERSEY

LIBRARY OF CONGRESS CATALOGING-IN-PUBLICATION DATA

PHILANDER, S. GEORGE.

IS THE TEMPERATURE RISING? : THE UNCERTAIN SCIENCE OF GLOBAL

WARMING / S. GEORGE PHILANDER.

P.   CM.

INCLUDES BIBLIOGRAPHICAL REFERENCES AND INDEX.

ISBN 0-691-05775-3 (CLOTH : ALK. PAPER)

1. GLOBAL WARMING.   2. ENVIRONMENTAL SCIENCES—PHILOSOPHY.

3. HUMAN ECOLOGY.   I. TITLE

QC981.8.G56P48   1998

551.5′2—DC21   97-37613   CIP

THIS BOOK HAS BEEN COMPOSED IN PALATINO TYPEFACE

THE VERSE ON P. 7 IS REPRINTED BY PERMISSION OF FABER AND FABER FROM "THE HOL-
LOW MEN," WHICH APPEARS IN T. S. ELIOT, *COLLECTED POEMS 1909–1962*, 1974.

PRINCETON UNIVERSITY PRESS BOOKS ARE PRINTED ON ACID-FREE PAPER

AND MEET THE GUIDELINES FOR PERMANENCE AND DURABILITY OF THF

COMMITTEE ON PRODUCTION GUIDELINES FOR BOOK LONGEVITY OF THE

COUNCIL ON LIBRARY RESOURCES

HTTP://PUP.PRINCETON.EDU

PRINTED IN THE UNITED STATES OF AMERICA

1 3 5 7 9 10 8 6 4 2

To My Father

If the Lord Almighty had consulted me before embarking on the Creation, I would have recommended something simpler.

Alfonso X of Castille
(Alfonso the Wise)

CONTENTS

PREFACE

W E ARE GAMBLING that the benefits of our industrial and agricultural activities—increasing standards of living for the rich and poor alike—will outweigh possible adverse consequences of an unfortunate by-product of our activities, an increase in the atmospheric concentration of greenhouse gases that could lead to global climate changes associated with global warming. Some experts are warning that we are making very poor bets; others assure us that the chances of global warming are so remote that the outcome will definitely be in our favor. We conduct opinion polls, but the matter remains unresolved because scientific differences, unlike political disagreements, cannot be settled by means of referenda. This impasse is disquieting because the issue is of vital importance to each of us; it concerns the habitability of our planet. To what extent are we interfering with the processes that maintain the benign conditions under which a glorious diversity of fauna and flora is flourishing? How should we proceed in the face of contradictory answers from the experts?

In our attempts to cope with this complex planet, we could learn from our success in coping with another immensely complex system, the human body. In the same way that public awareness of the manner in which the human body functions facilitates effective health care, so public awareness of the processes that maintain benign conditions on our remarkable planet will facilitate effective environmental policies. Earth's habitability is too important a matter to be left entirely to experts, especially when they contradict each other for reasons that are ideological rather than scientific. Everyone ought to participate in discussions of environmental policies and to that end should have a rudimentary understanding of the processes that make this a habitable planet.

Everyone already has considerable familiarity with the phenomena that contribute to Earth's habitability. They are the things people write poetry about: clouds, rain, wind, weather, the oceans. Although we find those phenomena endlessly fascinating and full of mystery, we nonetheless tend to view their scientific aspects in simplistic terms. For example, some people believe that summer is warmer than winter for the "obvious" reason that the Earth is closer to the sun in summer than in winter. They are reluctant to accept that very dense gases such as chlorofluorocarbons (CFCs) can rise into the stratosphere and harm the ozone layer. The most common and serious error is the assumption that scientists should always be capable of precise predictions. The expectation that more accurate scientific information will soon be available can result in a continual deferral of action to deal with environmental problems until there is a crisis. The likelihood of a calamity will decrease once everyone realizes that many natural phenomena, even seemingly mundane ones such as clouds and winds, are so complex that scientific results concerning them have inevitable uncertainties. To appreciate that scientists can nonetheless provide valuable information to help us cope better with our environmental problems, we need some familiarity with a new discipline known as the Geosciences or Earth Sciences.

The Geosciences integrate traditional disciplines such as geology, biology, meteorology, and oceanography in order to address scientific questions about Earth's habitability. This subject has such an enormous scope that any one book can deal at most with a few of its facets. This book focuses on Earth's climate and its sensitivity to perturbations, those that occurred in the past—that resulted in the Ice Ages, for example—and those we are currently introducing, which could cause global warming. The main goal is to give insight into the science of the intricate processes that make this planet habitable in order to shed light on controversial environmental issues. Part One, which concerns general issues, starts with a chapter examining the principal reason for controversies: uncertainties in scientific results that cause a blurring of the distinction between science and policy. Chapter 2 describes briefly how scientists explore complex phenomena by studying simpler, idealized situations to obtain results that can be of enormous practical value even when precise predictions are precluded. Part Two examines some of the physical and chemical processes that make this a habitable planet. The topics include interactions between light and air molecules that enable the atmosphere to be a shield that protects us from dangerous ultraviolet rays and also a blanket that keeps the surface warm by providing a greenhouse effect (chap. 3); the dependence of Earth's great diversity of climatic zones on the global redistribution of heat and moisture by means of convection and clouds (chaps. 4, 5), winds that range from gentle sea breezes to the mighty Jet Streams (chap. 6), and chaotic weather patterns (discussed in chapter 7 along with a description of computer models that predict weather and climate). Chapters 8 and 9 concern oceanic currents and interactions between the ocean and atmosphere that cause phenomena such as El Niño and La Niña. Part Three describes how the interplay between the various processes discussed in Part Two determines Earth's response to various perturbations: the intensification of sunlight over the past few billion years (chap. 10); the periodic fluctuations in the distribution and intensity of sunlight on Earth that cause the cycles of seasons and of Ice Ages (chap. 11); our introduction of CFCs into the atmosphere and the resulting ozone hole over Antarctica (chap. 12); and the current increase in the atmospheric concentration of greenhouse gases that could lead to global climate changes (chap. 13).

Henry David Thoreau cautioned that "men are never tired of hearing how far the wind carried men and women, but are bored if you give them a scientific account of it." This book nonetheless attempts to explain to laymen the fascinating science of phenomena associated with our weather and climate. It is based on notes I prepared for an introductory course I teach at Princeton University. To accommodate those with little affinity for mathematics, the main part of the text is essentially void of equations and should be accessible to anyone interested in weather, climate, and related environmental issues. The Appendixes are intended for those who use the book as a text for an introductory course. It assumes familiarity with simple algebra and revisits some of the scientific arguments of the previous three parts, providing technical details plus exercises and suggestions for further reading.

I owe thanks to many people. My home institutions, the Department of Geosciences of Princeton University and the Geophysical Fluid Dynamics Laboratory of NOAA, provide an ideal environment for the study of weather and climate. I am grateful to friends, colleagues, and students who shared their expertise, commented on the manuscript, and generously offered advice and encouragement. I am indebted to Leo Donner, Kevin Hamilton, Rob Hargraves, Peter Heaney, Isaac Held, Philippe Hisard, Gabriel Lau, George Mellor, and Lori Perliski for critical readings of various chapters. For insights into good writing and good pedagogy I thank Harriet Bryan, Jim, Karen, and Majory Wunsch. Dan Feiveson, Jessica Godfrey, Barbara Winter, and Cathy Raphael contributed the splendid figures and computer graphics. My research has been supported by NOAA (grant NA56GP0226) and NASA (contract UCLA-NASA-NAG5–2224). Dr. G. Kukla, the American Meteorological Society, Cambridge University Press, the University of Washington Press, and *Science* generously gave permission to reproduce certain figures.

PART
ONE

1

BETWEEN THE IDEA AND THE REALITY

WE ARE IN A RAFT, gliding down a river, toward a waterfall. We have a map but are uncertain of our location and hence are unsure of the distance to the waterfall. Some of us are getting nervous and wish to land immediately; others insist that we can continue safely for several more hours. A few are enjoying the ride so much that they deny that there is any imminent danger although the map clearly shows a waterfall. A debate ensues but even though the accelerating currents make it increasingly difficult to land safely, we fail to agree on an appropriate time to leave the river. How do we avoid a disaster?

To decide on appropriate action we have to address two questions: How far is the waterfall, and when should we get out of the water? The first is a scientific question; the second is not. The first question, in principle, has a definite, unambiguous answer. The second, which in effect is a political question, requires compromises. If we can distinguish clearly between the scientific and political aspects of the problem, we can focus on reaching a solution that is acceptable to all. Unfortunately, the distinction between science and politics can easily become blurred. This invariably happens when the scientific results have uncertainties.

Suppose that we have only approximate, not precise estimates of the distance to the waterfall. Rather than leave it at that—rather than accept that we can do no better than predict that we will arrive at the waterfall in thirty minutes plus or minus ten minutes—some people will minimize the distance and insist that we will arrive in twenty minutes or less, while others will maximize the distance, stating confidently that we won't be there for forty minutes or more. Do these people disagree for scientific reasons? (Some may have more confidence in their instruments than others do.) Or do their different opinions simply reflect the difference between optimists and pessimists?

To cope with this problem, we usually start by addressing the uncertainties in the scientific results. After all, everyone knows that science, in principle, can provide precise answers. One of the first scientists to be acclaimed by the public for his accurate predictions was Isaac Newton:

> Nature and Nature's law lay hid in night
> God said, Let Newton be! and all was light.
> (Alexander Pope, "Epistle XI. Intended for Sir Isaac Newton in
> Westminster Abbey")

Since Newton's accomplishments in the seventeenth century, scientists have continued to impress the public with remarkably accurate predictions that have led to inventions that continue to transform our daily lives. If, today, the results concerning a certain scientific problem have uncertainties, then, surely, it is only a matter of time before scientists present us with more accurate results. It is therefore easy to agree on a postponement of difficult political decisions regarding certain environmental problems on the grounds that we will soon have more precise scientific information. This could prove disastrous should we suddenly find ourselves at the edge of the waterfall. We recently had such an experience.

The current fisheries crisis, which is most severe off the shores of New England and eastern Canada where many species of fish have practically disappeared, started a decade after scientists first warned that overfishing could cause a dangerous reduction in fish stock. The scientists sounded a timely alert, but poor judgment on the part of policymakers contributed to this disaster. That is not how policymakers view the matter. Some complain of the scientists' "penchant for speaking in terms of probabilities and confidence intervals" and propose that, in future, scientists make "more confident forecasts . . . to catch the attention of regulators." As is often the case in environmental problems, we arrived at an impasse because of the reluctance of scientists to give definitive answers and the unwillingness of policymakers to make difficult political decisions. United States Congressman George Brown, former chairman of the House Committee on Science, Space and Technology, wonders whether there is a conspiracy between these two groups, the scientists who are assured a continuation of funds to improve their predictions, and the politicians who avoid difficult decisions that can cost them their jobs.

The fisheries crisis exemplifies a type of environmental problem with which we have had ample experience, and which the biologist Garret Hardin describes as "a tragedy of the commons."

> Picture a pasture open to all. It is to be expected that each herdsman will try to keep as many cattle as possible on the commons. Such an arrangement may work reasonably satisfactorily for centuries because

tribal wars, poaching, and disease keep the numbers of both men and beast well below the carrying capacity of the land. Finally, however, comes the day of reckoning, that is, the day when the long-desired goal of social stability becomes a reality. At this point, the inherent logic of the commons remorselessly generates tragedy.

As a rational being, each herdsman seeks to maximize his gain. Explicitly or implicitly, more or less consciously, he asks, "What is the utility to me of adding one more animal to my herd?"

The benefit of one more animal goes entirely to the herdsman. When it is sold, he receives all the proceeds. The disadvantage, the additional overgrazing, is shared by all. It is clearly to the advantage of the herdsman to acquire another animal. The other herdsmen reason similarly. The result is ruin for all.

The creation of private ownership is one attempt to avoid a tragedy of the commons. The landowner, out of self-interest, will prevent the land from being ruined. His interests do not necessarily coincide with ours so that we place restrictions on some of his actions. For example, he has to observe regulations concerning the disposal of sewage and toxic wastes because the water below his land and the air above it, fluids that can move pollutants off his property, remain part of the commons.

Given that, in the past, we successfully avoided many tragedies of the commons, why did we fail to avoid a fisheries crisis? Part of the reason is the novelty of the phenomenon; a decline in fish stock on a global scale is without precedent (although we have decreased the whale population significantly). Whereas we readily accept regulations that minimize damage that might occur during disasters with which we have experience (e.g., building codes that maximize public safety during an earthquake), we often oppose regulations that amount to precautionary measures to mitigate potential environmental disasters for which there are no precedents. If such disasters should occur in relatively small regions, they will serve as painful lessons on the need for regulations. If, however, a potential disaster has a global scale, we cannot afford to learn our lesson in such an expensive manner. Finding ways to avoid global disasters is a matter of urgency because the rapid growth in our numbers, and in our technological prowess, is increasing the likelihood of such disasters.

The English curate Thomas Robert Malthus (1766–1834) anticipated some of the problems that are likely because of the steady rise in the human population. In 1798 he predicted that, because our numbers are increasing at a rate that far exceeds the rate at which arable land increases, we are heading for a "gigantic inevitable famine." His forecast proved wrong, at least in the case of Britain and other rich countries, because he failed to anticipate the extent to which scientific and technological advances would increase the productivity of the inhabitants of those countries. The rising standards of living in the rich countries led to social changes that decreased the number of children born per family, thus stabilizing the populations. Presumably, the poor nations, by raising their standard of living, will in due course also halt the growth of their populations. Perhaps the present rapid rise in the world population is a temporary phenomenon, to be followed by a period of declining populations, whereafter the world population will stabilize at a relatively low number that our planet can accommodate comfortably. We all wish for such an end but, unless we are careful, the journey could prove very treacherous. We will face serious problems should the poor nations copy the current industrial and agricultural practices of the rich because, at present, the cost of a high standard of living is an enormous, adverse impact on the environment. The damage has been reversed, or at least mitigated in a few cases—certain rivers, once so polluted that they occasionally caught fire, are now clean and safe for fish—but other escalating environmental problems go essentially unattended. The fisheries crisis is one example. Another worrisome development is the rapid accumulation of greenhouse gases in the atmosphere. Rich countries may have limited the rate at which their populations grow, but they are increasing the rate at which they inject greenhouse gases into the atmosphere.

Toward the end of the nineteenth century, the Swedish chemist Svante August Arrhenius (1859–1927) alerted the world that our industrial activities, which are causing the increase in the atmospheric concentration of greenhouse gases, could result in global climatic changes. Nobody paid much attention to his predictions because of considerable uncertainties. For example, in the absence of instruments with which to monitor atmospheric carbon dioxide levels, many scientists assumed that oceanic absorption of that gas would prevent its accumulation in the atmosphere. During the past century, scientists have reduced the uncertainties significantly. There is now indisputable evidence that the atmospheric concentrations of several greenhouse gases, not only carbon dioxide, have indeed been increasing rapidly since the start of the Industrial Revolution. Mathematical models of Earth's climate now provide details of the global climate changes, including global warming, that we should expect. Furthermore, recent studies of past climates on Earth, which tell us about the response of this planet to perturbations, enable us to gauge the likely consequences of the perturbations that we are introducing. Empirical and theoretical evidence (reviewed in chap. 13) leave no doubt that the growth in the atmospheric concentration of greenhouse gases, if continued indefinitely, will cause global climatic changes. There is, however, considerable disagreement about the timing of those changes. Some experts paint alarming pictures of sea level that will soon rise to inundate New York, London, Tokyo, and other coastal cities; of pests and diseases that will spread into new territory; and of fertile farmlands that will soon become drought-stricken. Other experts assure us that our industrial and agricultural activities pose no immediate threat, that there is no likelihood of global warming in the foreseeable future. Do these contradictory statements reflect uncertainties in the scientific results, or are they expressions of ideological differences? Here we have another example of an impasse created by uncertainties in scientific results, and a reluctance to make difficult political decisions. The difficulty stems from our reluctance to accept that, although accurate predictions are, in principle, possible on the basis of the laws of physics, such forecasts may be impossible in practice because scientists—especially those who study complex environmental problems—deal with idealizations of reality. They too have to accept that

> Between the idea
> and the reality
> Between the motion
> and the act
> Falls the Shadow
>
> *(T. S. Eliot, "The Hollow Men")*

During the century since Arrhenius first sounded an alert, scientists have decreased the uncertainties in his forecasts considerably and are likely to continue doing so. However, there will always be shadows cast by inevitable uncertainties. We therefore have to ask ourselves whether we can continue to defer action much longer, given that the problem we face is similar to that of the gardener in the following riddle.

A gardener finds that his pond has one lily pad on a certain day, two the next day, four the subsequent day and so on. After 100 days the pond is completely filled with lily pads. On what day was the pond half full?

Answer: Day 99

Suppose that the gardener, once he realizes what is happening, quickly enlarges the pond to twice its size. On what day will the new pond be completely filled?

Answer: Day 101

The riddle illustrates how any problem involving explosive growth requires action at a very early stage, long before there are clear indications of impending trouble. In the case of the debate about global warming, in which some people insist that we are close to day 1 while others are adamant that we are close to day 100, the riddle indicates that, far more important than a precise answer that brings the debate to an end, is recognition of the special nature of the problem, its geometric growth. With such problems, it is far wiser to act sooner rather than later. To defer action is to court disaster.

A major impediment to progress on novel environmental problems, such as global warming or the depletion of fish stock, is the unrealistic expectation of precise predictions endorsed unanimously by the scientific community. This expectation reflects ignorance of the trial-and-error methods by which scientists reduce uncertainties in their results. Scientists continually subject any proposed solution to tests and do not hesitate to modify (or even abandon) a solution should it prove inadequate. Sound scientific results that have logic and clarity as their hallmark are often achieved by making many missteps along a tortuous road. (The irony is similar to that of poets who labor arduously to produce poems that flow effortlessly.) In our attempts to cope with our environmental problems, we should adopt a similar approach of trial and error. Rather than implement comprehensive programs that decree a rigid course of action to reach grand, final solutions, we should promote adaptive programs whose evolution is determined by the results from those programs and by new scientific results that become available. It will then be easier to take action when there is no scientific consensus, and it will be possible to correct mistakes at an early stage before scarce resources have been wasted. By adopting this approach, we are doing remarkably well in our efforts to minimize damage to Earth's protective ozone layer.

Because they recognize that the atmosphere is a commons whose protection is their responsibility, the nations of the world agreed in the Montreal Protocol of 1987 that each would limit its production of the chlorofluorocarbons (CFCs) that contribute to the depletion of the ozone layer. This was a remarkable decision because it was made before there was clear evidence that CFCs are harmful to the ozone layer; at the time, scientists had only warned that CFCs could pose a serious threat. The diplomats who negotiated the Montreal Protocol accepted the uncertainties in the scientific predictions and proceeded to take action. They wisely agreed on regulations that are subject to periodic reviews in order to accommodate new scientific results. The initial regulations called for a reduction in the production of CFCs. When the original predictions concerning the effect of CFCs on the atmosphere proved erroneous—scientists at first underestimated the harmful effects of CFCs (see chap. 12 for details)—the regulations were made more stringent, and the nations decided to cease production of CFCs.

Progress in science depends on the continual testing of results and explanations. Such skepticism makes it highly unlikely that scientists will ever unanimously recommend a solution to a problem that is so complex that the results have inevitable uncertainties. For a specific problem, the available evidence at a certain time may favor one particular explanation—e.g., overfishing for the disappearance of fish—but because of uncertainties, other possibilities—such as poor sampling of the fish population—cannot be excluded. A continual refinement of measurement and theories reduces uncertainties causing the spectrum of scientific opinions to converge. As long as there is some uncertainty, however, a few dissenting voices will persist. These contrariants, although they are wrong most of the time, are valuable because they force a continual reexamination of scientific methods and results. On a few rare occasions, they are even right. A prominent example concerns the idea of continental drift. With the exception of a few dissenters, the geological community rejected this notion for many decades, but in the end the dissenters proved right. Today the majority of geologists accept that continents drift.

The evidence accumulated over the past 100 years—especially the rapid scientific progress over the past few years—has convinced most scientists that the current rapid increase in the atmospheric concentration of greenhouse gases will lead to global climatic changes. There are, of course, a few dissenters, who would probably be skeptical even if the scientific issues were of strictly academic interest and concerned another planet, Mars, for example. That the issues are not strictly of academic interest but also have political aspects complicates matters enormously and dramatically alters the role of the skeptics, who become the focus of considerable attention for reasons unrelated to the merits of their scientific arguments. By focusing attention on the small group of dissenters, those who wish a continual deferral of action create the false impression that there is little agreement in the scientific community. To appreciate what is happening, the public needs to become familiar with the methods and results of scientists, especially the reasons for inevitable uncertainties that preclude precise predictions with which everyone agrees.

Scientists can contribute to the mitigation of potential disasters even when they are unable to make precise predictions. Consider the case of earthquakes. Their time of occurrence cannot be predicted, but it is possible to anticipate how Earth's surface will move should an earthquake occur and hence to build structures capable of surviving earthquakes. To ensure public safety, states enforce building codes that are in accord with the recommendations of earthquake engineers. The public, familiar with the disasters that earthquakes can cause, readily accepts those regulations. We need to recognize the need for regulations even in the case of environmental disasters for which there are no precedents. To avoid disasters such as the depletion of fish stock off the northeastern coast of the United States, we can demand of scientists more confident forecasts that "catch the attention of the regulators," but it would be wiser to accept that we have to act in spite of uncertainties, in spite of the inevitable shadow between the idea and the reality.

We cope successfully with some environmental problems but not with others. We sensibly agreed to limit the release of CFCs into the atmosphere, but we failed to act in time to avoid a fisheries crisis. We have yet to do something about the accumulation of greenhouse gases in the atmosphere. While we inject those gases into the atmosphere at an accelerating rate, we defer a decision on how soon to make a transition to environmentally sound technologies because of uncertainties in the scientific predictions and even bigger uncertainties about the cost of the transition. We are rushing toward dangerous rapids and possibly a waterfall but are reluctant to act because we do not know precisely how much time we have left before we are in serious trouble. In discussions about the appropriate time to leave the river, we should keep in mind that a step as drastic as leaving the river promptly and trekking across unknown terrain is but one option. It may be wiser to start by leaving the swift, accelerating part of the stream and moving where the flow is slow. Coping with uncertainties is not a novel challenge. All of us—businessmen, politicians, military strategists—routinely make decisions on the basis of uncertain information, usually after we have familiarized ourselves with the available facts. We who are privileged to live on this benign planet should at least attempt to understand it so that we can assess the likely consequences of our actions.

2

IS OUR PLANET FRAGILE OR ROBUST?

IN THE BEGINNING, swarms of rocks and swirls of gas circled the Sun. Gravity, the force that attracts objects to each other, gradually transformed this stony rubbish into something rich and strange: nine sparkling planets that wander across the skies as if they were independent of the stars. All are wondrous worlds—Saturn is adorned with rings, Jupiter with several moons—but only Earth, one of the smaller and less spectacular planets, a fragile blue dot when seen from afar, is blessed with a miracle, a glorious diversity of flora and fauna. Only our planet is habitable.

Earth has been habitable for most of its long history, even in its youth when the Sun was far fainter than it is today. At the birth of the solar system, some 4.5 billion years ago, the intensity of sunlight was approximately 30% less than at present. If the Sun were suddenly to become as faint as it originally was, temperatures on Earth would drop so much that all the water would freeze. The geological record nonetheless indicates that our planet has had plentiful water in liquid form, and hence has maintained a moderate range of temperatures, practically since birth. This paradox of the faint Sun but warm Earth—a clear indication that ours is a resilient planet, capable of maintaining benign conditions in spite of adversities—is a blessing to our species, *Homo sapiens*, because today we thrive on conditions that could have evolved only on a planet that has favored life for billions of years. It is as if elaborate preparations preceded the advent of mankind.

We are the beneficiaries of gradual evolutionary processes that unfolded to the rhythms of cyclic phenomena including the repeated buildup and erosion of mountains and the periodic opening and closing of ocean basins. Today our varied landscape, its spectacular mountains, vast plains, lush tropical jungles, and barren deserts, is but a snapshot of a continually changing panorama. *Homo sapiens* has been present for such a brief period, a million years or so, that we have witnessed only one possible arrangement of the continents. Our predecessors, however, have seen many changes because life, in one form or another, has been present on this planet practically since its birth.

Earth's inhabitants, far from being passive guests, have influenced conditions on this planet throughout its history. We are indebted to earlier life-forms for contributions that range from the oxygen in the air we breathe to the fossil fuels on which our civilizations have become dependent. During the evolution of the conditions that suit us so well, many species became extinct. The geologic record provides ample evidence of catastrophic extinctions of numerous species on several occasions. Some probably contributed to their own demise. Certain primitive forms of life that thrived in Earth's original atmosphere without oxygen produced oxygen as a waste, to such an extent that an enormous amount accumulated in the atmosphere. We, too, could be causing ourselves considerable inconvenience.

We have been here for but a blink of an eye on geologic time scales, but, because our development has been favored by easy access to fuels, minerals, and metals, we have grown in numbers and in prowess with incredible rapidity. We have become an important geologic agent and are in the process of interfering with the processes that make this planet habitable. For example, we are modifying the composition of the global atmosphere significantly because of the fossil fuels we are burning at a furious rate. Soon our actions will cause the atmospheric concentration of greenhouse gases to increase, not by a tiny percentage, but by a factor of two at least.

It is unlikely that the biosphere as a whole will be endangered by our actions; it has survived bigger calamities in the past. The global warming that we are likely to cause has been exceeded in earlier epochs. The present cold era our planet has been experiencing for some two million years was preceded by a period during which the poles were free of ice. At different times in the past, temperatures have been much higher and much lower than they are today, but because of fortuitous factors such as our distance from the Sun and the size of Earth, temperature extremes on this planet never approached those that prevail on our neighbors Venus and Mars. In the long run, these factors will enable our planet to continue maintaining habitable conditions; our actions will be of little consequence over the next thousands and millions of years.

That is scant comfort to humans, however, because we are vulnerable to even modest climate changes that persist for only a few years or decades. We *Homo sapiens* can ill afford something as trivial as an increase in the frequency of hurricanes, or prolonged droughts in some regions and repeated devastating floods in others. Our planet may seem robust from the perspective of the entire biosphere—life has been on Earth for more than three billion years—but it can nonetheless appear fragile from the perspective of individual species, especially us. That is why there is cause for concern about the global environmental consequences of our agricultural and industrial practices. The matter has generated lively debates, but they often end in stalemates; for example, some insist that Earth is robust, others that it is fragile. In reality it is both!

Stalemates in debates about environmental issues frequently stem from a failure to appreciate how enormously complex our planet is and how that complexity leads to inevitable uncertainties in scientific results concerning the consequences of our activities. Familiarity with the geosciences, and especially with the methods by which scientists arrive at their results, will contribute to more constructive debates.

Scientists, poets, and painters all explore the natural phenomena that make this a habitable planet. What is distinctive about the perspective that scientists provide? Consider their response when they are asked what matter is. They reply that it is composed of molecules. Molecules, in turn, are made of atoms. Atoms consist of electrons, protons, and neutrons. The two latter "elementary" particles are made of even more "elementary" particles, an ever lengthening list of exotic fermions, baryons, bosons, quarks, etc. Apparently, science is a form of poetry that describes in terms of metaphors without providing true explanations.

These remarks have an element of truth because science explains in terms of certain fundamental laws that govern all natural phenomena. The classical Greeks were the first to propose that there are governing laws behind nature's great variety of particular occurrences. This idea was probably a projection onto the universe of the orderly life of the Greek polis, where a citizen enjoyed considerable freedom—from the whims of a ruler, for example—even though his life was rigorously bound by impersonal laws. Nature would be less arbitrary and capricious than it seems if it, too, were governed by laws. Scientists have found that the laws of nature are far more absolute than civil laws. Whereas the latter can be disobeyed, with some inconvenience perhaps, those of nature must be obeyed; there is no choice. The development of this idea into the sciences of the past few centuries would astonish even the ancient Greeks.

Scientists use the governing laws of nature to make inferences and predictions. They progressed very slowly during the Middle Ages when natural philosophers relied strictly on logical deductions from self-evident truths—their version of the governing laws of nature—to explain natural phenomena. Those early scientists were reluctant to make measurements to check their inferences because they believed that the fallible impressions of the senses are untrustworthy. (Some contemporaries of Galileo's declined to look through his telescope for fear of sullying their minds with false impressions.) The upheavals of the Renaissance and Reformation gradually changed attitudes and set the stage for the experimental methods of modern science. The transition was marked by the invention of many instruments: the microscope, thermometer, and pendulum clock were invented late in the sixteenth century, the telescope and barometer early in the next century. News of these instruments, and of the measurements made with them, spread rapidly across Europe because of journals that scientists started to publish to promote communication among themselves and to generate interest in their activities. In Britain, the Royal Society started to publish the *Philosophical Transactions: Giving some Accompt of the Present Undertakings, Studies and Labours, of the Ingenious in many Considerable Parts of The World.* The motto of the Royal Society, *Nullius in Verba*, has been translated as "take nobody's word for it; see for yourself." The key difference between the investigations of scientists and those of poets and painters is an insistence on measurements to check the validity of scientific arguments, especially the governing physical laws. To illustrate the rule of measurements in science, we next summarize the efforts of early scientists to gain an understanding of the phenomenon of heat.

Some of the first experimental studies of heat tried to determine whether its opposite, cold, has similar properties. For example, can the flow of "cold" from an object such as a block of ice cool its surroundings the way the flow of heat from a hot object warms its surroundings? After numerous experiments, scientists concluded that ice, or any cold object, cools its surroundings, not because it emits "cold," but because heat flows from the warmer to the colder objects the way water flows downhill. This conclusion led to the conjecture that heat is a type of fluid, without mass or color, that always flows from warmer to colder objects. (Some regarded it as the soul of matter, called *phlogiston*, which escapes from or is absorbed by an object when it is burned.) Careful experiments, especially by the French chemist Lavoisier, seemed to reveal that this fluid, later named caloric fluid, can neither be created nor destroyed. If it leaves a warm object it is bound to reappear elsewhere, in a colder one. If it is known how much caloric fluid the warmer object loses, then it is possible to calculate accurately by how much the colder one warms up. The law that caloric fluid is conserved—which enabled scientists to explain a host of phenomena involving the transfer of heat and even enabled them to improve the efficiency of the heat engine, the workhorse of the

Industrial Revolution—is a particular case of a type of natural law that scientists invoke repeatedly to explain various phenomena. The ancient Greek philosopher Democritus first articulated such a law when he stated, around 400 B.C., that "nothing can arise out of nothing; nothing can be reduced to nothing." In other words, all apparent change is but the rearrangement of unchanging parts; nothing comes into being, or perishes, in the absolute sense of the words. (Democritus identified atoms as invariants, as the fundamental building blocks of all matter; the Greek word atom means "indivisible.")

Our world may be changing continually, but it nonetheless has invariants; it obeys conservation laws. For a while scientists believed caloric fluid to be an invariant. That theory had glorious successes, but it also had some embarrassing flaws. For example, there is a simple experiment in which caloric fluid seems to be created out of nothing: if we rub our hands together, they get warm. Where does the heat come from? Why does the law for the conservation of caloric fluid fail?

One of the first gentlemen to address this problem was the versatile and peripatetic American, Benjamin Thompson, from Rumford (now Concord) in New Hampshire. Knighted by George III of England, and named Count Rumford by a duke of Bavaria, this soldier, diplomat, and gardener—he laid out Munich's famous *Englischer Garten*—was also an inventor (of special baking ovens, for example) and a scientist. The following passage is from an essay Count Rumford read before the Royal Society of London in 1798.

It frequently happens that in the ordinary affairs and occupations of life opportunities present themselves of contemplating some of the most curious operations of Nature; and very interesting philosophical experiments might often be made, almost without trouble or expense, by means of machinery contrived for the mere mechanical purpose of the arts and manufactures . . .

Being engaged, lately, in superintending the boring of cannon, in the workshops of the military arsenal at Munich, I was struck with the very considerable degree of Heat which a brass gun acquires, in a short time, in being bored; and with the still more intense Heat (much greater than that of boiling water as I found by experiment) of the metallic chips separated from it by the borer.

The more I meditated on these phenomena, the more they appeared to me to be curious and interesting. A thorough investigation of them seemed even to give a farther insight into the hidden nature of Heat; and to enable us to form some reasonable conjectures respecting the existence or non-existence of an igneous fluid: a subject on which the opinions of philosopher have, in all ages, been much divided . . .

From whence comes the Heat actually produced in the mechanical operation above mentioned? Is it furnished by the metallic chips which are separated by the borer from the solid mass of metal?

Further experiments by Rumford, and subsequent ones by James Joule in the 1840s, established that friction, and more generally motion, can generate heat. This meant that heat cannot be a material substance, that the caloric theory of heat must be wrong. (Rumford, who helped topple the theory that Lavoisier proposed, married Lavoisier's widow after her husband lost his head to the guillotine.)

In their search for an alternative to the caloric theory of heat, scientists turned to the venerable Democritus and adopted his proposal that atoms are the building blocks of all matter. They proposed that atoms, and arrangements of atoms called molecules, are in a continual state of random, disorganized motion. In a solid, the atoms are tightly bound together and vibrate randomly, back and forth, while remaining in essentially one position. In a liquid, they have more freedom and can move relative to one another. In a gas, molecules are entirely free to move in any direction whatsoever, spinning and rolling and changing direction whenever they collide with each other. Scientists no longer considered heat a substance (fluid) but now regarded it as something far more abstract; they associated it with the random motion of molecules and atoms. This led to the recognition that there are three ways in which heat can flow from a warm to a cold body. Conduction requires physical contact between the warm and cold bodies because the most energetic molecules literally nudge their immediate neighbors into less lethargic states. Convection occurs in fluids when a warm parcel consisting of a huge number of energetic molecules moves through the fluid from one region to another. Radiation involves the vibrations of atoms, and especially their electrons, at the surface of a body. Such a surface can radiate heat across a vacuum. That is how heat from the Sun reaches Earth.

The molecular theory of matter led to the recognition that heat is but one form of energy and that it can readily be exchanged for other forms of energy. During such exchanges—when we rub our hands together, friction causes the conversion of kinetic energy into thermal energy—total energy, not heat, remains invariant. The conservation of energy is accepted as a universal law of nature to this day.

For a while, it was not evident that the theory associating heat with the wild motion of mysterious, invisible molecules, is preferable to the caloric theory, which regards heat as a mysterious, invisible fluid. Many scientists objected to the lack of direct evidence for the existence of molecules that move randomly. They were not persuaded until, early in the twentieth century, Albert Einstein provided convincing evidence by explaining how the random motion of minuscule dust particles suspended in transparent water is caused by collisions between the dust and the much smaller, randomly moving water molecules. (Einstein was twenty-six years old and an examiner at the Swiss Patent Office in Berne when he published his paper on the random motion of dust particles. It was but one of three groundbreaking papers he published in 1905 in *Annalen der Physik*. The two other papers concerned the photoelectric effect [see chap. 3] and the theory of relativity. The latter work brought Einstein so much fame that few people realize that he was a physicist of the first rank even without the theory of relativity.)

Scientific understanding—of heat, for example—comes from an insistence on measurements to check the validity of scientific arguments, especially the validity of any proposed laws that govern natural phenomena. The test for any law is the accuracy of predictions made by means of that law. Isaac Newton, one of the early members of the Royal Society, formulated the laws that govern the motion of inert bodies, from molecules to planets and comets, and used them to predict the motion of the planets with astonishing precision. His successors discovered laws that govern phenomena not known to him, those associated with electricity and magnetism and with the atomic structure of matter, for example. Today many scientists believe that they are on the verge of establishing all the laws that govern natural phenomena. Some claim that they will then have a "Theory of Everything." As far as Earth's weather and climate are concerned, we already know all the governing laws and, in principle, should be capable of precise predictions. Why is that not possible in practice?

Phenomena that can be studied by means of controlled, replicable experiments can often be predicted with remarkable accuracy. The results have been translated into impressive feats that range from colossal machines that fly smoothly through the air, to electronic marvels that transform our daily lives. A physicist at a university in southern California introduces his students to this class of phenomena by starting his series of lectures with a dramatic demonstration. He takes a large, heavy metal ball, suspended from the high ceiling by means of a taut cable, into his hands and steps backwards until the ball just touches his nose. He then releases the ball, now a pendulum, so that it swings from him until it reaches the end of its arc. Next the ball starts to accelerate back toward the physicist. Everybody grows tense when he remains absolutely motionless, but, to the relief of all, the ball comes to a halt just short of his nose.

This demonstration of confidence in the accuracy of the predictions that can be made on the basis of the laws of physics has an air of magic and never fails to impress. It is a very effective introduction to certain branches of science usually referred to as the "hard" sciences but, unfortunately, it also contributes to the delusion that precise prediction is the hallmark of all science. That belief is evident when some people respond to inaccuracies in weather forecasts by declaring that weather prediction is "an art, not a science"; they are reluctant to accept that many phenomena are so complex that precise predictions are impossible. For an appropriate introduction to the latter phenomena, which form the focus of the "soft" sciences—fields such as biology, ecology, geology, psychology—the story of the professor with the pendulum should be modified as follows.

The professor teaches at a university in southern California, a region of frequent tremors and earthquakes. Suppose that a minor tremor, which causes no damage, occurs during the pendulum demonstration. The professor, a man of common sense despite his flair for miracles, will promptly take shelter under the nearest table. When he finally emerges, he could complain about the chance occurrence that ruined his beautiful demonstration. He would, however, be far wiser to regard the chance occurrence as a splendid opportunity to explore fascinating phenomena, like earthquakes, which do not lend themselves to controlled experiments. Why are earthquakes more common in some regions than others? Why is the Pacific rim, a region prone to earthquakes, also a "ring of fire" with many active volcanoes? How does Earth's surface move during earthquakes? Answers to these questions enable us to mitigate the disasters associated with earthquakes, even if we are unable to predict exactly when they will occur.

To understand earthquakes, we have to recognize that Earth, in addition to sunlight, has another important source of energy, heat generated by the radioactive decay of certain elements deep below Earth's surface. Energy from the Sun sets the atmosphere and ocean in motion. Heat in Earth's interior generates motion whose surface manifestation is the very slow drift of continents that incites sudden earthquakes and violent volcanic eruptions. Our planet is alive! It spins about its tilted axis, orbits the Sun, moves its atmosphere and oceans in chaotic patterns, and punctuates the stately drift of its continents with sporadic quakes and eruptions. The challenge is to explain how this frenzy of activity contributes to the maintenance of conditions that suit us so well.

To address questions concerning the habitability of Earth, scientists engage in an interplay between measurements and theory, but the approach is different from the one used in studies that involve replicable experiments. In studies of complex phenomena, measurements serve, not so much to test the validity of certain laws of nature, as to understand the roles that different processes play in determining those phenomena. This can be accomplished by applying the known laws of physics to a hierarchy of idealized versions of reality in order to obtain results that can be compared to measurements. For example, to explore Earth's habitability, we start by idealizing our planet as a perfect sphere, the same size as, and as far from the Sun as Earth. Given the intensity of the incident sunlight, it is then possible to calculate the temperature of this model of Earth by invoking the law for the conservation of energy. This exercise amounts to a test of the hypothesis that our planet has reasonable temperatures because it happens to be at the right distance from the Sun. The hypothesis is valid if the calculated temperature agrees with what is measured.

The surface of the model planet turns out to be far colder than that of Earth. It does not follow that the law for the conservation of energy is invalid. Rather, our model is too idealized and needs additional features; it needs to acquire an atmosphere that serves as a blanket. (The atmosphere needs to provide a greenhouse effect.) We are in effect refining our original hypothesis: temperatures at Earth's surface are determined not only by its distance from the sun but also by Earth's atmosphere. The new model is again idealized by assuming that the atmosphere is a static layer of gases without winds or weather. Once more we appeal to the laws of physics to calculate a temperature for the surface of our model planet. This time the temperature is too high. The model needs further refinements.

We seem to be in the process of building a habitable planet. We start with a very simple model, a solid, rotating sphere without any atmosphere. When it proves to be deficient, too cool, we add a blanket, a static atmosphere. Now the surface of the model planet is too hot. To remedy this problem, the model must acquire an ocean and winds that evaporate moisture from the ocean, thus cooling it. The winds take heat from the ocean primarily in low latitudes and transport that heat to higher latitudes, thus preventing the tropics from being too hot and the polar regions from being too cold. This global redistribution of heat and also of moisture involves phenomena such as chaotic weather patterns and meandering oceanic currents, which are driven by the winds. A model planet with these features—an atmosphere and an ocean, both in motion, is habitable; it has moderate temperatures, running water (a hydrological cycle), and a great diversity of climatic zones.

Next, for the planet to be robust in the face of an intensification of sunlight over billions of years, the greenhouse effect of the atmosphere needs to be varied by continually changing the composition of the atmosphere. This can be accomplished by accelerating or decelerating the recycling of gases and other chemicals between the solid Earth and its fluid envelopes, the oceans and atmosphere. Volcanic eruptions are spectacular examples of the transfer of material from the solid Earth to the atmosphere. The return of that material to the interior of Earth depends on the slow drift of the continents, which is the surface manifestation of turbulent motion deep in the hot interior of our planet. The maintenance of that motion over billions of years requires that the size of a planet exceed a certain minimum (otherwise it will have insufficient radioactive material to maintain high temperatures). Earth, of course, satisfies that requirement. It has had epochs during which volcanic eruptions were far more frequent than they are today. At such times the atmosphere had a different composition—it had more carbon dioxide—so that an amplified greenhouse effect could counter the faintness of the sun (see chap. 11)

The beauty of geoscience is its integration of seemingly disparate disciplines—geology, seismology, volcanology, oceanography, meteorology—in order to explain why ours is a special planet on which life has evolved. The salient result from this new field is that our well-being at the surface of Earth depends on an enormously complex interplay of fascinating phenomena and processes. It is essential that we appreciate this complexity because failure to do so can result in the adoption of simplistic approaches to difficult problems, and in the unrealistic expectation of precise predictions from environmental scientists. Our planet is so intricate that many of its phenomena, such as weather, have very limited predictability. This introduces an apparent paradox: if we are unable to forecast the weather more than a few days hence, how can we hope to anticipate global climate several decades from now?

In the 1950s the meteorologist Edward Lorenz, in an effort to explore the limited predictability of weather, launched a new branch of science, popularly referred to as the study of chaos. This field deals with the complex behavior of seemingly simple systems that can be described in terms of precise mathematical equations. A particularly simple example is a wallet that falls from the pocket of a skier and slides down a ski slope. The path that the wallet travels is complicated because of the slope's innumerable little hills and valleys shown in figure 2.1. These moguls, as they are known, can deflect the wallet sideways as it slides downhill. To calculate their effect, it is necessary to solve the equations that describe Newton's laws, subject to given

Figure 2.1 Moguls on a ski slope. A skier is barely discernible in the upper part of the figure. From Lorenz (1993).

"initial conditions." This term refers to the wallet's initial position, speed, and direction when it first falls onto the snow. In principle there should be no problem in determining where, at the bottom of the slope, the skier should look for his wallet. In practice, there are significant difficulties. Figure 2.2. shows seven different trajectories for a wallet that, each time, starts with exactly the same speed and direction; only its initial position changes, by a mere millimeter, each time. At first these differences are so slight that they have an insignificant effect on the path of the wallet. By the time the wallet has traveled 60 m (180 feet) downhill, however, the differences have grown enormously. The paths continue to diverge rapidly thereafter. It follows that, in practice, it is impossible to predict where the skier will find his wallet if it slides too far downhill.

The results in figure 2.2 were obtained not for an actual wallet sliding down a realistic slope such as the one in figure 2.1, but for an extremely idealized and simplified case. In the calculations, the moguls were the regular, idealized ones in figure 2.3, and the wallet, rather than an object with a complicated shape, was an object shrunk to the size of a pinhead! In spite of these considerable simplifications, the path of the wallet is predictable for only a very limited time! These results are humbling because, in principle, there is no problem. It is a matter of solving certain exact equations subject to initial conditions. Difficulties arise when we try to determine those initial conditions; innumerable chance occurrences make it impossible to determine exactly where the wallet falls onto the snow, certainly not to within a fraction of a millimeter. The tiny yet unavoidable errors are the ones that ruin the predictions. Phenomena such as the sliding wallet are said to demonstrate extreme sensitivity to initial conditions. The prediction of the weather several days hence, which requires a description of the state of the atmosphere today, is similarly limited by inaccuracies in the description of that initial state. In picturesque terms, failure to take into account every butterfly that flaps its wings limits the predictability of weather.

Results from the study of chaos shed glaring light on the severe limitations of scientists, their limited ability to predict something as simple as the path of a wallet sliding down a ski slope. In principle, there is no problem; accurate predictions are possible on the basis of the laws of physics. In practice there are immense difficulties. Should we conclude that, given our difficulties with the simple wallet, there is little point in attempting forecasts of the response of our immensely complex planet to perturbations such as an increase in the concentration of greenhouse gases?

Scientists studying chaos have discovered that there often is method

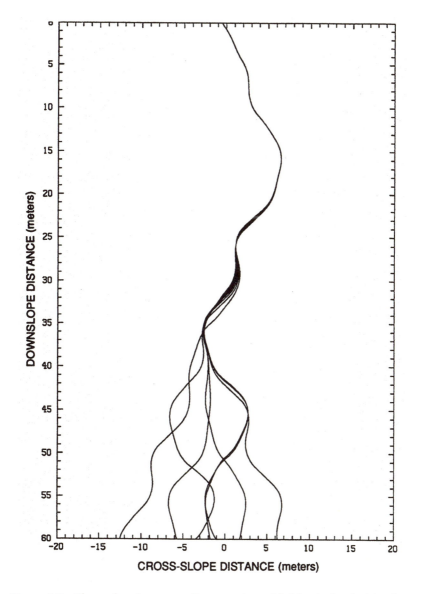

Figure 2.2   The paths of seven wallets, starting with identical velocities from points spaced at intervals of 1 mm along a west-east line. From Lorenz (1993).

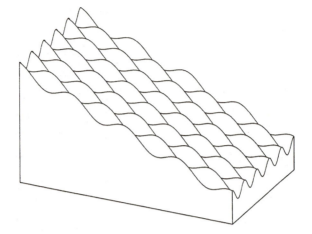

Figure 2.3 An oblique view of a section of the model ski slope. From Lorenz (1993).

to the madness of seemingly chaotic phenomena. For example, in the case of the wallet sliding down the ski slope, they have found that the path has certain predictable aspects. (See Appendix 2 for a reference that provides further information.)

We may have very limited ability to predict, on the first day of January, say, what the weather will be on a specific day a week later, let alone a specific day in July, but we can nonetheless forecast, in January, that an average day in July will be much warmer than an average day in January. This climate change between January and July, a consequence of the steady increase in the intensity of sunshine during that period, affects certain aspects of weather in a predictable manner. To explore the relation between weather and climate, we return to the simple pendulum.

Consider a pendulum in the form of a heavy metal ball suspended from the ceiling by a cable. Assume that the ball is so heavy that it gradually stretches the cable. A long pendulum is more ponderous than a short one so that the period of oscillation of the pendulum gradually increases. Assume that this change, which is perfectly predictable provided the ceiling is stationary, is from 1 to 10 seconds in the course of a month. If we happen to be in a region of earthquakes, and if random tremors occur continually, then the pendulum's period of oscillation will be unpredictable at a specific time, but there will nonetheless be a discernible change over a month. Initially the times it takes for the completion of successive oscillations will be clustered around 1 second:

.5 sec; 1.2 sec; .8 sec; 1.3 sec; 1.2 sec . . .

but after a month, the oscillations of the much longer pendulum will be clustered around 10 seconds:

9.0 sec; 12.1 sec; 11.1 sec; 8.5 sec; 11.5 sec . . .

In other words, although it is impossible to predict how the pendulum will move from one oscillation to the next, it is possible to predict how it will move on the average as its length increases. The pendulum has a natural mode of oscillation, back and forth at a high frequency, and also has forced variability, the gradual lengthening of the cable. The latter affects the former. In the atmosphere, the forced variability, climate, similarly influences the natural variability, weather, which is why weather in summer is different from that in winter.

A confident prediction, in the heart of summer, that temperatures will be dramatically lower a few months later, in winter, surprises no one. The seasonal change in sunlight is known to be large, and we know from experience that Earth's response to that change is large. Far more astonishing is the prediction that, over the next tens of thousands of years, Earth will cool dramatically as an Ice Age develops (assuming that man does not interfere). This prediction can be made with confidence because the geological record clearly shows that a similar phenomenon has occurred repeatedly in the past. The recurrent Ice Ages (discussed in chap. 11) are the response of our planet to periodic changes in the distribution of sunlight caused by slight variations in factors such as the tilt of Earth's axis and the eccentricity of its orbit. This response is remarkable for having an enormous magnitude even though its cause, a change in sunshine, is very modest. It clearly is possible to anticipate certain aspects of how our immensely complex planet will respond to perturbations.

Scientists are now developing computer models of Earth's climate in order to simulate and explain present and past climates and to anticipate the global climate changes that are likely in the future because of the current rapid rise in the atmospheric concentrations of greenhouse gases. The models, which are rapidly growing in realism, provide valuable information about future conditions and can contribute to the mitigation of possible disasters. Thus far, results from the models have generated heated debates, in part because Earth's rich spectrum of natural variability (e.g., weather) is superimposed on the forced variations in response to seasonal changes in sunshine and to the steady increase in the atmospheric concentration of greenhouse gases. Our planet's spontaneous music can mask the forced changes.

The transition from winter to summer can be difficult to pinpoint because of erratic weather fluctuations. Detecting climate changes in response to higher atmospheric carbon dioxide levels can similarly be problematic because of climate fluctuations such as a few exceptionally hot summers or a succession of mild winters. This is part of the reason for the current debate about global warming, the topic of chapter 13; scientists agree that a continual increase in the atmospheric concentration of greenhouse gases will in due course lead to global warming but disagree about the timing of those changes. Climate models provide useful information about future climate changes that are likely, but the models do have uncertainties.

The Ice Ages were in response to very modest changes in the distribution of sunlight. The lessons to be learned from this result are both encouraging and discouraging. On the one hand, the result implies that it should be possible to anticipate Earth's response to modest perturbations. On the other hand, it is disconcerting to learn that Earth's response to slight disturbances can be astonishingly severe!

The geological record of the Ice Ages teaches us to be apprehensive about the considerable disturbance that we are creating by means of our industrial and agricultural activities. By increasing the atmospheric concentration of greenhouse gases significantly we are meddling with very intricate processes that maintain benign conditions at the surface of this planet. The models that scientists are developing to predict the consequences of our actions are growing in realism and provide valuable information for policymakers. However, every model is an idealization; no model can anticipate every possibility. We may need insurance against unanticipated disasters. This message emerges clearly from the story of the ozone hole over Antarctica; it is a timely, cautionary tale.

In the 1970s, three atmospheric chemists—Mario Molina, Sherwood Rowland, and Paul Crutzen—in research for which they were later awarded a Nobel Prize, correctly anticipated that certain chemicals that we release into the atmosphere, CFCs, would harm the ozone layer in the stratosphere. Their alert enabled governments to take precautionary measures. These steps proved to be prudent because it was soon recognized that the magnitude of the problem had been underestimated. Scientists had predicted a very gradual loss of ozone over several decades; instead, a gaping hole suddenly started to appear in the ozone shield over Antarctica every October. The phenomenon is so bizarre that, if it had been predicted, the forecasts would probably have been rejected as implausible. The ozone hole could be explained only with hindsight. It is similarly possible that scientists will misjudge some of the consequences of the increasing amounts of greenhouse gases that we inject into the atmosphere. Uncertainties are inevitable. To make prudent decisions, we need to familiarize ourselves with the available information concerning the processes that make this planet habitable and the sensitivity of these processes to perturbations. The following chapters describe some of those processes. It will become evident that in the long run, over tens of thousands of years, we are unlikely to do great harm to our planet as a whole. We can, however, cause our particular species considerable inconvenience over the next several decades by perpetuating practices that continually amplify the perturbations we are introducing.

PART
TWO

3

LIGHT AND AIR

OUR ATMOSPHERE serves us in several capacities: as a parasol for shade from sunlight that is too bright, as a shield for protection from dangerous ultraviolet rays in sunlight, and as a blanket that traps heat and thus keeps Earth's surface comfortably warm. The atmosphere succeeds in those functions even though it is merely a thin veil of gases; if our planet were the size of an apple, its atmosphere would be as thick as the apple peel. (Some 80% of the mass of the atmosphere is within 10 km of the surface of the Earth, which has a diameter exceeding 12,000 km.) Our comfort and safety depend on intricate interactions between photons of light and the molecules of that thin layer of gases. Throughout Earth's history, those interactions have maintained temperatures sufficiently moderate to ensure plentiful water in liquid form, a condition vital for the evolution of life. Conditions are very different on Venus and Mars, our inhospitable planetary neighbors. Temperatures are so low on Mars that all its water is frozen, and they are so high on Venus that its water evaporated and escaped to space long ago. Venus is closer to the Sun than we are; Mars is farther away. At first it may appear that we are special because we are at the right distance from the Sun. There is, however, far more to the story. The composition of the atmosphere of a planet is another very impotant factor; it is one of the main reasons why Earth's surface is safe and warm, whereas that of Venus is uncomfortably hot. (Mars is cold because it has almost no atmosphere as measured by the pressure at its surface.) To understand the importance of the composition of an atmosphere, we have to explore the nature of light and its interactions with air molecules.

Light, a topic redolent of religion and mysticism, retains much of its mystery even after innumerable scientific investigations. The first experiments were the remarkably simple ones by the great Isaac Newton who proposed that a beam of light is a stream of discrete particles. Next, the Dutchman Huygens argued that Newton was wrong, that a beam of light is a continuous, undulating wave. Various nineteenth century experiments confirmed Huygens's views, but twentieth-century scientists decided that both Huygens and Newton were right! Scientists have come to the uneasy conclusion that light has a strange, dual character: a beam of light is both a smoothly undulating train of waves and a lumpy stream of discrete photons. For an infinitesimally small photon to resemble a wave, which is spread over a region in space, the photon must be capable of being in several places at the same time! Such bizarre assumptions have led to the invention of the electronic marvels that are changing our daily lives. They also enable scientists to explain the various ways in which enigmatic photons of light interact with insubstantial molecules of air to produce fascinating phenomena that range from rainbows, blue skies, and red sunsets, to the greenhouse effect that depends, not on the most abundant gases in our atmosphere, nitrogen and oxygen, but on trace gases that are present in minute amounts. The following discussion of these interactions covers topics that range from the classical physics of Newton, to the electromagnetic theory of Maxwell, to the quantum theory of Planck and Einstein.

Earth, Mars, and Venus Compared

Temperatures at the surface of Venus are so high that lead would melt there; on Earth they vary over a moderate range that suits us well; on Mars, temperatures are so low that whatever water it has is frozen. The three solid black dots on figure 3.1 correspond to the actual temperatures at the surfaces of these planets. To what degree are the different temperatures attributable to different distances from the Sun? The continuous curve in figure 3.1 provides the answer; it shows how the temperature of a planet would decrease with increasing distance from the Sun if the planet absorbed all its incident sunlight. To obtain that curve (see Appendix 3.4 for details), we invoke the law for the conservation of energy by assuming that, in a state of equilibrium, there is a balance between the amount of heat that a planet absorbs and the amount it radiates to space. Because the planets are isolated objects in a vacuum, space, and can gain or lose heat only by means of radiation, the only form of energy that comes into play is radiational heat. (At the surfaces of the terrestrial planets, their internal sources of heat are negligible in comparison with the heat received from the Sun.) From measurements of the intensity of sunlight at Earth's surface, it is possible to calculate how much heat Earth absorbs. We then know how much heat Earth radiates into space in a state of equilibrium. This information can be converted into a temperature for Earth because the heat a surface radiates increases as its temperature rises. (Toward the end of the nineteenth century, the Ger-

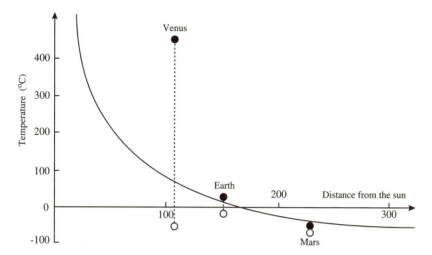

Figure 3.1 The curve shows the decrease in temperature, with increasing distance from the Sun (in units of 10^6 km), for planets that absorb all the incident sunlight and that have neither internal sources of energy nor atmospheres. The open circles take into account that each planet reflects some sunlight. The solid circles correspond to the actual temperatures at the surfaces of the planets. The length of each dashed line is a measure of the greenhouse effect.

man physicists Stefan and Boltzman established the mathematical formula that relates the temperature of a surface to the amount of heat that it radiates.) Temperatures for the other planets can be calculated similarly as explained in Appendix 3.4.

The results in figure 3.1 reveal discrepancies between the calculations (the continuous curve) and the measurements (the solid black dots), especially in the case of Venus which is far hotter than the calculations would lead us to expect. The discrepancies get worse if we make the calculations more realistic by taking into account that, of the sunlight that is incident on a planet, only a portion is absorbed because some is reflected to space. (The planets and the moon are visible, not because of the heat they radiate, but because of the sunlight they reflect.) The reflectivity of a surface, its albedo, depends primarily on its brightness. The brighter a color, the more it reflects and the less it absorbs. (We therefore tend to wear light colors in summer, dark colors in winter.) Earth, primarily because of its white clouds, reflects some 30% of its incident sunlight. (Table 3.1 lists the albedo of different surfaces on Earth.) Venus is even more extravagant. She sparkles in the evening sky because she is draped in a thick

TABLE 3.1
Albedo (or reflectivity) of Various Surfaces (percent)

Bare soil	10–25
Sand, desert	25–40
Grass	15–25
Forest	10–20
Snow (clean, dry)	75–95
Snow (wet and/or dirty)	25–75
Sea surface (sun > 25° above horizon)	<10
Sea surface (low sun angle)	10–70

layer of clouds that reflects 75% of the incident sunlight. What is not reflected is absorbed; therefore, Earth absorbs 70% of the incident sunlight, Venus only 25%. Venus absorbs less heat than does the Earth even though Venus is much closer to the Sun. If we take this factor into account in calculating the temperature of the planets, then the results correspond to the open circles in figure 3.1; the planets are even colder than the continuous curve in figure 3.1 indicates. The low temperatures on Mars can now be explained as a consequence of its albedo and distance from the Sun. Earth and Venus, however, pose a puzzle: why are they so warm? Furthermore, how is it possible for Venus to be so much hotter than Earth when it actually absorbs less sunlight than does Earth? To answer these questions we have to take into account the atmospheres of the planets that can act like blankets that trap heat.

Consider identical twins, one wrapped in a thin sheet, the other in a thick blanket. If they are outside where it is very cold, then the first will be chilly and the second comfortably warm, even though their bodies produce exactly the same amount of heat. That heat flows from the body of each twin, through the blankets, to the cold air that surrounds them. The thin sheet offers practically no resistance to the flow of heat but the blanket offers considerable resistance. For the loss of heat to be the same in the two cases, the thickness of the blanket has to be countered by a large temperature difference across the blanket. (The rate of heat loss through the blanket decreases as the blanket gets thicker but increases as the temperature difference across the blanket gets larger.) That is why it is warm beneath a thick blanket that effectively resists the flow of heat, but cold underneath a thin sheet that is poor at inhibiting the flow of heat.

The atmosphere of a planet can act as a blanket that traps heat. (It is said to provide a greenhouse effect.) Temperatures are, of course, higher beneath a blanket than above it. We therefore have to distinguish between the temperatures at the surface of a planet and those aloft, in the upper atmosphere. The measurements (black dots) in figure 3.1 correspond to temperatures at the surfaces of the planets; the open circles, which are based on the entirely correct argument that a planet in a state of equilibrium radiates as much heat as it absorbs, correspond to the cooler conditions in the upper atmosphere. The differences, the dotted lines, provide measures of how "thick" the planetary atmospheres are. The atmosphere of Venus is seen to amount to a thick eiderdown, that of Mars to a gossamer thin sheet. Our atmosphere has just the right thickness for life on this planet. In the absence of our atmosphere, the average temperature at Earth's surface would be a frigid −18°C, all water would be frozen, and we would probably not be here. The greenhouse effect of our atmosphere increases surface temperatures to +15°C and contributes enormously to the habitability of our planet.

The effectiveness with which an atmosphere intercepts heat from the surface of a planet depends on its composition, its concentration of greenhouse gases. Venus has a huge greenhouse effect because its atmosphere is composed primarily of carbon dioxide, a very effective greenhouse gas. Mars, too, has an atmosphere composed primarily of carbon dioxide, but it has very little atmosphere. The surface pressure, a measure of how much atmosphere a planet has, is 1000 mb on Earth, only 6 mb on Mars. The small size of Mars is part of the reason why it has so little atmosphere. Some scientists believe that because the force of gravity is so small on Mars, it could have lost much of its atmosphere during the impact of a large meteor.

The most abundant gases in Earth's atmosphere, nitrogen and oxygen, are unable to absorb infrared heat from the Earth's surface. Together they account for more than 99% of our atmosphere so that our modest but vitally important greenhouse effect depends on gases, primarily carbon dioxide, water vapor, and methane, that are present in very small amounts. Carbon dioxide accounts for only 0.035% of our atmosphere. (Table 3.2 shows the composition of our air at ground level.) This is such a small amount that we, because of our current industrial and agricultural activities, are in the process of increasing the atmospheric concentration of carbon dioxide significantly. If maintained, this increase is bound to cause global warming.

Our atmospheric blanket has very unusual properties: although it traps heat from the surface of Earth, it is transparent to heat from the sun, our main source of heat. What is the difference between heat from the sun and that from the surface of a planet? Additional questions about the reasons for different conditions on different planets concern the albedo of a planet. Earth has a high albedo because of its white clouds, which reflect a large fraction of the incident sunlight. If the clouds are composed of colorless water droplets, why are they

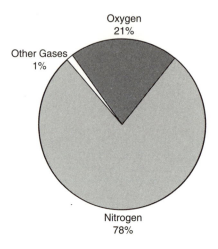

TABLE 3.2

Composition of Dry Air at Ground Level in Remote Continental Areas

Constituent	Formula	Concentrations (%)
Nitrogen	N_2	78.1
Oxygen	O_2	20.9
Argon	Ar	0.93
Carbon dioxide	CO_2	0.035
Neon	Ne	0.0018
Helium	He	0.0005
Methane	CH_4	0.00017
Krypton	Kr	0.00011
Hydrogen	H_2	0.00005
Ozone	O_3	0.000001–0.000004

white, and hence highly reflective? To address these questions, we have to explore the nature of light.

The Scattering of Light

In 1666, at the age of 23, Isaac Newton bought a glass prism "to try therewith the phenomena of colors." He describes his experiment as follows: "In a very dark Chamber, at a round Hole, about one third Part of an inch broad, made in the Shut of a Window, I placed a Glass Prism, whereby the Beam of the Sun's Light, which came in at that Hole, might be refracted upwards toward the opposite Wall of the Chamber, and there form a colored Image of the Sun." Newton next devised an experiment in which a ray of a single color passed through a second prism, and discovered that there was no further dispersion; the color remained unchanged. He concluded that "the Sun's light is an heterogeneous Mixture of rays." He next confirmed that all colors are the components of white by using a lens to bring the complete spectrum of colors to a common focus. The colors disappeared to produce white light. Newton concluded that the color of any body is simply the color in sunlight that the body reflects the most.

Newton regarded a beam of light as a stream of tiny particles, corpuscles. This view probably stemmed from his enormous success in explaining the motion of particles, including the orbits of planets. Light reflects off a surface the way a ball bounces off a wall. When it passes from air into water or glass it refracts the way the path of a stream of particles is bent when the speed of each particle changes upon entering another medium. Such phenomena can readily be explained by appealing to Newton's most important work, *Principia Mathematica*, in which he formulated the laws that govern the motion of all particles. But if a beam of light is a stream of particles, how can two beams cross without interfering with each other? Such considerations led the Dutch physicist Christian Huygens to propose that a beam of light is a continuous train of waves. Newton disagreed because waves do not always travel in straight lines the way light does, and do not leave sharp shadows behind objects the way light does. (He overlooked that extremely short waves create a sharp shadow behind a large object). Although Huygens succeeded in explaining a number of optical phenomena in terms of the wave theory, Newton's authority was such that the corpuscular theory prevailed for almost a century after his death. It was abandoned in 1800 when experiments by the Frenchman Augustin Jean Fresnel and the Englishman Thomas Young firmly established the wave nature of light (fig. 3.2).

In one of Young's experiments, light enters a dark room, not through one small hole in the "shut" of the window as in Newton's original experiment, but through two holes (fig. 3.3). When the holes are relatively large, then the light passing through them creates two bright spots on the screen some distance away. However, when the holes are very small, dark rings appear in the overlapping bright spots on the screen. The dark rings can be explained by assuming that waves spread outward from each of the holes. In the dark areas of the screen, the crests of the waves from one hole arrive simultaneously with the troughs of the waves from the other hole. The cancellation of crests by troughs causes darkness. In the bright areas on the screen, crests from one hole reinforce crests from the other hole. The wavelength of the light and the distance between the hole and the screen determine the location of the light and dark rings. The agreement

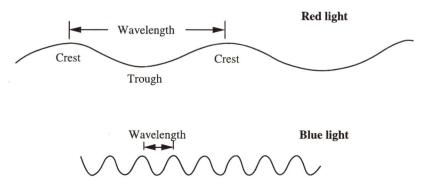

Figure 3.2 In the wave theory of light, the distance between successive crests (or troughs) is known as the wavelength. In the case of visible light, a difference in wavelength corresponds to a difference in color; waves of red light are long; those of blue light are short.

between the calculations and the measurements established the wave nature of light and vindicated Huygens.

Huygens and his followers established that a beam of light is a continuous train of waves. The distance between successive crests, the wavelength, determines the color of the light: red light has a long wavelength; blue light, a short wavelength. Sunlight is the superposition of all the waves or colors in the rainbow. This quantitative ap-

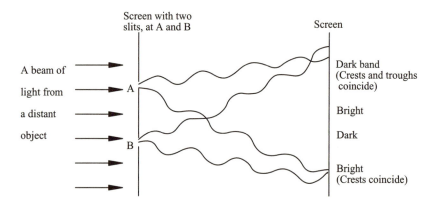

Figure 3.3 In Young's experiment, light from a distant source passes through two narrow slits, A and B, to reach a screen. Light from the two slits form two distinct bright bands if A and B are far apart, but form interference patterns of alternating light and dark bands if A and B are sufficiently close. In the latter case, the crests of waves from A and B reinforce each other in the bright bands. In the dark bands the crests of waves from one slit and the troughs from the other slit cancel each other.

proach makes scientists seem dry as dust in comparison with poets and painters who associate colors with emotions: green for envy, red for passion, blue for coolness. The beauty of the scientific approach emerges from the diverse and seemingly unrelated phenomena that can readily be explained if we accept that different colors correspond to different wavelengths. Why, for example, are clouds white? Why is the sun a yellow disk in a blue sky when seen from Earth, but a white disk in a black sky when seen from the Moon? To answer these questions, we must investigate how tiny particles affect a beam of light. This is a moot matter in the case of the Moon, which has practically no atmosphere and hence no suspended particles to interfere with sunlight. On the Moon, the color of the Sun is the sum of the rainbow colors, white, and the sky has no color because there is no light coming from it. Stars are visible even when the Sun is shining. The Earth is different because it has an atmosphere that is composed of molecules and suspended particles that scatter light. The rule that governs scattering is simple if the size of a particle is very small relative to the wavelength of the light: the smaller the particle, the less effect it has on the wave. (A particle with zero size has no effect.) Alternatively, the effect of a small particle of a fixed size decreases as the wave becomes longer. The tiny gas molecules of which the atmosphere is composed are negligibly small in comparison with long red waves but are sizable in comparison with short blue waves. They readily scatter blue waves but have little effect on red waves, which is why the sky is blue. On a sunny day, each part of the sky scatters blue waves toward us. When we look directly at the Sun, we see beams that have lost so much of their blueness that the color of the Sun corresponds to the color of the longer-surviving waves. That color is primarily yellow when the Sun is overhead and becomes progressively redder as the Sun moves toward the horizon, and the sunbeams traverse a thicker and thicker layer of atmosphere. When only the very long red waves succeed in reaching us we have a red sunset or sunrise. The color is particularly intense after a volcanic eruption that deposits into the atmosphere particles sufficiently large to scatter all but the longest wavelengths.

In addition to tiny gas molecules, the atmosphere contains aerosols, which are suspended solid or liquid matter such as dust particles, water droplets, and ice crystals. Those aerosols, whose sizes can greatly exceed the wavelength of any color in sunlight, have a scattering effect that is independent of wavelength. In other words, they scatter all the colors in sunlight and therefore appear white. An isolated large particle may not scatter much light—a water droplet, for example, permits most of the light that falls on it to travel right through it—but an aggregate of such particles, water droplets in a cloud or frozen ones in snow, scatters a significant amount of light and appears white. For this same reason, the sky is hazy white when the atmosphere has a considerable concentration of aerosols or pollutants with a size large enough to scatter all the colors in sunlight. The amount of aerosols in the atmosphere has been increasing because of industrial activities, which also increase the atmospheric concentration of greenhouse gases. Whereas the greenhouse gases contribute to global warming, the aerosols contribute to cooling because they reflect sunlight. This could be one reason why, over the past century, the increase in atmospheric carbon dioxide has not been accompanied by a steady rise in temperatures.

A raindrop reflects some of the sunshine that falls on it, but most of the light penetrates the drop and is refracted, a term that refers to the bending of a beam so that its direction changes (fig. 3.4). The drop is in effect a prism because the different colors in sunlight are bent to different degrees. This can result in a rainbow, a beam of sunlight refracted into its constituent colors as in Newton's experiment. To see a rainbow, we must position ourselves carefully: our backs must be to the Sun, and we need to face a distant rainshower or the spray from a garden hose.

Clouds at great altitudes are so cold that they are composed of ice crystals rather than raindrops. The crystals often are hexagonal in shape and can refract sunlight or moonlight to create a halo. In principle, the halo can be colored like a rainbow, but it usually is not. The considerable variations in the sizes and shapes of ice crystals cause overlapping bands of color so that the result is a white halo. The first sign of an approaching storm is often a very thin layer of extremely high cirrus clouds composed of ice crystals. A ring around the moon can therefore be an omen of foul weather.

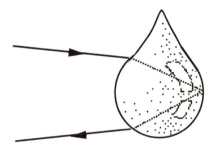

Figure 3.4 A rainbow, which can appear when we stand with our backs to the Sun, involves the refraction of sunlight when it enters a water droplet, its reflection off an inside surface of the droplet, and its refraction toward us when it leaves the droplet.

The Radiation of Light

Ours is the blue planet because that is the principal visible color, along with white swirls of cloud, that Earth reflects to space. In addition, Earth radiates invisible infrared light to space. Whereas sunlight readily travels through Earth's atmosphere to reach the surface, much of the infrared light from the surface is absorbed by the atmosphere, which acts as a blanket. The resultant greenhouse effect depends on the different wavelengths of radiation from the Sun and that from Earth, a consequence of the different temperatures of the Sun and Earth. To observe how the luminosity of an object changes as its temperature changes, watch the heating unit of an electric kitchen range that has been turned on in a dark room. When lukewarm, the heating unit emits radiant but no visible light. When its temperature reaches 600°C, it has a reddish glow. The color changes to cherry red and then bright orange or yellow with increasing temperatures. Ultimately, the iron melts. Higher temperatures are possible if, instead of iron, we heat tungsten wire enclosed in an inert atmosphere, in other words, if we heat the filament of an electric bulb. At 2000°C, tungsten emits a yellowish light. The Sun, at 6000°C, has light with a good deal of blue, as can readily be confirmed by consulting a painter or by passing the light through a prism. A pattern is becoming apparent: as the temperature increases, the dominant color systematically moves through the spectrum of the rainbow, from red to yellow to blue. Figure 3.5 shows how the range of colors (wavelengths) that a body emits changes to shorter and shorter wavelengths as its temperature increases.

The hotter a body, the shorter the wavelength of its dominant color. The very hot Sun emits predominantly short, yellow waves. The much cooler Earth emits long infrared waves. For an atmosphere to provide a greenhouse effect, it must therefore distinguish between the very short waves in visible light and the long infrared waves from Earth's surface. To appreciate that different types of radiation, despite these differences, have much in common, we must turn to two seemingly unrelated topics, electricity and magnetism.

Magnets are fascinating because they can reach across empty space, and even through a thin sheet of paper, to attract metals some distance away. A bar of amber that has been rubbed and electrified can similarly attract certain tiny objects that are electrically charged. Electrical and magnetic forces have many similarities, and both are reminiscent of the gravitational force that one body exerts on another, the Sun on Earth, for example. In all three cases, the various forces decrease in exactly the same way with increasing distance, as the in-

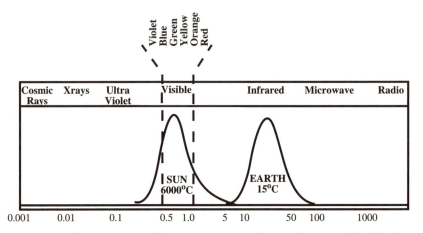

Figure 3.5   The spectrum of electromagnetic radiation from two blackbodies with temperatures of 6000°C, that of the Sun, and 15°C, that of the much cooler Earth. The wavelength, indicated along the horizontal axis, is measured in microns: 1000 microns = 1 mm. Visible light, the colors of the rainbow, occupies the small part of the spectrum between the dashed lines. The vertical axis measures the intensity of radiation, except that the increase in the intensity of radiation as temperature increases—the difference between Earth and the Sun—has been suppressed.

verse of the square of the distance. It is reasonable to inquire whether the three forces—electrical, magnetic, and gravitational—are related. The first two are indeed related—as has been known since the nineteenth century—but the link to gravity, a problem that occupied Einstein for many years, is yet to be established.

In 1820 the Danish physicist Hans Christian Oersted discovered that the flow of electrical charges, an electrical current, through a wire causes that wire to become a magnet so that a compass needle orients itself perpendicular to the wire. Two parallel wires with electrical currents flowing in the same direction attract each other the way magnets do (fig. 3.6). Oersted's experiments established that an electrical current generates magnetism. Demonstrating the converse, that magnetism can create electricity, proved more difficult—numerous experiments with wires around magnets failed to produce a spark—until the remarkable Englishman Michael Faraday succeeded.

Faraday used magnetism to generate electricity in a coil of wire by moving a magnet back and forth through the coil. The critical factor was the movement of the magnet; the current was absent unless the magnet moved. If a fluctuating magnetic field can generate an electric

Figure 3.6 Electrical cables attract each other if their currents are flowing in the same direction, but repel each other if the currents are flowing in opposite directions.

current, which in turn creates magnetism then, Faraday reasoned, oscillating magnetic and electric fields should some how be able to push each other along to propagate through space like a disembodied wave. He lacked the mathematical training to pursue the idea, but it inspired James Clerk Maxwell, the great Scottish mathematician and physicist, to investigate the matter. In 1864 Maxwell demonstrated that oscillating electric and magnetic fields can indeed push each other along, and he calculated the speed of propagation of this effect, this electromagnetic wave, to be 3×10^{10} cm/sec (or 186,000 miles per second). This also happens to be the speed of light which had been measured many years earlier. Maxwell's equations furthermore provided detailed explanations for various phenomena in optics. Maxwell concluded that light corresponds to an electromagnetic wave.

The most important result from Maxwell's theory is that visible light waves represent only a small fraction of the spectrum of electromagnetic waves. The spectrum, which is shown in figure 3.5, ranges from long radio waves to micro and infrared waves, to visible light, to very short ultraviolet and x rays. The unity behind this diversity of phenomena is breathtaking. They all are electromagnetic waves; they travel at exactly the same speed, that of light; and they differ primarily in the distance between their successive crests, their wavelength. These waves can be generated by vibrating electrical charges. The frequency of the vibrations determines the frequency with which wave crests pass a fixed point. The higher the frequency, the greater the number of crests that pass a point, which implies a smaller distance between crests and hence a shorter wavelength. In 1886 the German physicist Heinrich Hertz succeeded in generating very long electromagnetic waves by means of electrical currents that oscillated at a low frequency. Shortly afterward, the Italian engineer Guglielmo Marconi invented the wireless telegraph that uses long radio waves to send messages over huge distances.

Vibrating electrical charges produce electromagnetic radiation. If the charges are the electrons in the atoms that constitute a solid, then the radiation has a spectrum of wavelengths, and we perceive it as light or heat. (Each atom consists of electrons spinning around a nucleus, not unlike planets orbiting the Sun.) In the case of a gas whose molecules, and hence groups of atoms, are isolated from one another, the radiation takes a simpler form; it corresponds not to a continuous spectrum but to a few isolated wavelengths or colors. For example, the gas potassium emits a distinctive red color; the gas sodium, a yellow color. It is as if the atoms in each freely moving molecule of a gas behave like a tuning fork that produces only a distinctive note. (The same chemical, in solid form, behaves like an assemblage of tuning forks; when tied together, they produce a broad continuous spectrum of notes. A heated solid similarly produces a continuous spectrum of colors.) Each gas has a distinctive spectrum so that chemicals in gaseous form can be identified from their spectra, the colors they radiate when heated. In the nineteenth century, the German physicist Gustav Kirchoff discovered that gases absorb the same colors that they emit. When light with a continuous spectrum—light from a white hot body, for example—is passed through a heated gas, that gas absorbs the colors it is known to emit so that black lines appear in the previously continuous spectrum. If the gas is sodium, for example, then a black line appears where the color yellow ought to be. Astronomers exploit this result to determine, from the missing lines in the spectra of light from the Sun and other stars, what the chemical composition of those heavenly bodies are.

At a temperature near 6000°C, the Sun emits a continuous spectrum of light, most of which is visible to the human eye. Certain gases in the outer layers of the Sun, the chromosphere, absorb some of the colors in sunlight and introduce thin black lines into the spectrum. When sunlight next travels through Earth's atmosphere, it encounters gases that absorb additional colors. Of particular importance to us is the absorption of ultraviolet rays by the gas ozone. (We return to this matter shortly). The sunlight that reaches Earth's surface heats it, causing it to glow and to radiate long infrared waves into space. In passing through our atmosphere, this earthlight encounters gases that absorb some of its wavelengths. Those are known as greenhouse gases, which contribute to global warming. To explain why some gases absorb ultraviolet beams in sunlight and others absorb infrared earthlight, we have to discuss ideas which, early in the twentieth century, shook the foundations of physics.

The Absorption of Light

Toward the end of the nineteenth century, it became clear that the physics of Newton and his successors could not provide explanations for phenomena that are encountered in extreme situations: at very high speeds close to that of light; at very high temperatures; and at very small scales, the scales of molecules and atoms. To explain the curious phenomena that occur at high speeds, Albert Einstein proposed the Theory of Relativity in which perceptions of space and time are relative and depend on the observer, whereas the speed of light remains an absolute constant. Of more concern to us in our attempt to understand the interplay of light and air is the more radical Quantum Theory which Max Planck proposed in 1899 when he addressed the discrepancy between theories and measurements regarding the colors that very hot bodies radiate. As the temperature of a body increases, the dominant colors of the radiation change from longer to shorter wavelengths. The theories could not explain this change. To resolve the problem, Planck hypothesized that electromagnetic radiation, which had been viewed as a continuous train of waves, actually consists of individual energy packages (or quanta) with well-defined amounts of energy per package. Previously, we associated different colors with different wavelengths. Planck proposed that different colors correspond to different quanta of energy; the longer a wave, the smaller the quanta. Thus, long red waves have relatively small quanta of energy; short blue waves have larger quanta. These assumptions enabled Planck to reconcile theory with the measurements, but his ideas did not gain wide acceptance until Einstein used them to explain the photoelectric effect. This effect refers to the following phenomenon: when a beam of light falls on a metal plate, it can generate an electrical current that flows from that plate to a nearby one, even in the absence of a connecting wire between the two plates. The light causes the atoms at the surface of the first plate to vibrate, so energetically that electrons break free and flow to the second plate, thereby establishing an electric current. Experiments show that this effect depends far more on the color of the light than on its intensity. For example, yellow light shining on copper produces almost no photoelectric effect, but even weak ultraviolet light on copper can generate a current. Einstein explained this curious result by reviving Newton's idea that light is a stream of discrete particles now called photons and by adopting Planck's theory that the shorter the wavelength of the light, the more energetic the photons. It is convenient to think of their energy in terms of money that is available in fixed denominations. Red light comes in units of $1 bills, more energetic yellow light in units of $5 bills, and very energetic ultraviolet light in units of $10. If we move beyond the visible spectrum to very short x rays we encounter $100 bills. In the other direction, very long radio waves have quanta of energy that correspond to pennies.

To explain the photoelectric effect—to explain how light interacts with matter—Einstein assumed that the energy of photons is available in discrete units and that different materials can absorb only certain discrete units of energy. For example, copper accepts the $10 bills of ultraviolet light but not the $5 bills of yellow light or the $1 bills of red light. Ten photons of red light, or two photons of yellow light may be the equivalent of $10, but copper refuses those denominations and insists on a $10 bill. In other words, an intense beam of yellow light cannot induce the photoelectric effect, only a beam of ultraviolet light can. Einstein's success in explaining the photoelectric effect earned him a Nobel Prize.

The photoelectric effect indicates that light is a stream of photons. But what about Young's experiment in which bright and dark rings appear on a screen when light passes through two small holes? That result leads to the conclusion that light is a continuous train of waves with crests and troughs. Let us return to the experiment in which light passes through a single slit in a screen. The illumination of the screen is uniform in a band opposite the slit. We can explain why the band can be wider than the slit by assuming either that waves spread outward from the slit or that photons near the edge of the slit are deflected slightly. Next let us turn to light that passes through two slits, as in Young's experiment. Now darker and lighter bands appear on the screen. From the particle point of view, this is unexpected because, with two slits, twice as many photons can reach the screen. Why is it that they fail to reach certain parts of the screen, the dark areas, which they had been able to reach when there was only one slit? For a photon that is passing through one slit to avoid a certain region when another slit some distance away is open, the photon has to know whether or not the other slit is open. How does it get this information? Is it possible for the photon to be at two different places at the same time? If we try to settle the last two questions by making measurements, we run into difficulties because, when we observe an object, we necessarily interfere with that object. For example, when we watch the trajectory of a tennis ball, we rely on photons of light to bounce off the ball and to travel toward our eyes. The ball is so heavy (it has so much momentum) and the photons are so light that the motion of the ball is practically unaffected by the photons that bounce off it. Matters are very different when we try to observe one photon by means of another. A collision is likely to deflect both photons. We can minimize the deflection by using, for observational purposes, photons of very long waves because the energy of a photon decreases as its wavelength increases. However, because a long wave is spread over a considerable distance, we are uncertain about the position of the photon. For accuracy in position, we need short wavelengths, but those are associated with high energies, something we want to avoid. There appears to be an intrinsic uncertainty in any attempts to make measurements.

In 1926 the Austrian physicist Erwin Schrodinger proposed an equation that describes the motion of matter in a manner that accepts uncertainty. Whereas the solution to Newton's equations tells us exactly where a particle will be at a certain time, the solution to Schrodinger's equation tells us the probability of finding a particle in a certain place. In the case of electrons that spin around the nucleus of an atom, the equation does not yield an orbit that is a specific circle or ellipse, but rather gives a "cloud" of "probability density" that tells us where the electron is most likely to be found. Although quantum mechanics, as this branch of physics is called, permits explanations for numerous phenomena, many people, find aspects of the theory troublesome. In 1926 Einstein, who found the intrinsic uncertainty particularly bothersome, wrote to a colleague: "Quantum mechanics is very impressive. But an inner voice tells me that it is not yet the real thing. The theory produces a good deal but hardly brings us closer to the secret of the Old One. I am at all events convinced that he does not play dice." Today, the success of quantum theory has been such that very few scientists share the view that it is "not yet the real thing" and that it will be replaced by something superior.

The shorter a wave, the more energetic are its photons. Short x rays are far more energetic than ultraviolet photons, which in turn make infrared photons seem lethargic. When a photon interacts with or is absorbed by matter, the consequences depend very much on how energetic the photon is. X rays are dangerous because they are powerful enough to change the chemical properties of an atom by knocking its electrons free. (As mentioned earlier, each atom consists of electrons spinning around a tiny nucleus, not unlike planets orbiting the Sun.)

The effect of a photon on an atom or molecule depends on the level of energy of the photon as shown in figure 3.7. The top panel shows how a very energetic ultraviolet photon can knock an electron free, in which case the atom is said to be ionized. Fortunately, such energetic photons are absorbed very high in the atmosphere, at altitudes greater than 60 km, in the ionosphere. The next panels in figure 3.7 show slightly less energetic, but still dangerous photons, that penetrate to lower altitudes where they knock electrons from one orbit to another—or photodissociate molecules—breaking them apart into their constituent atoms.

The photons that finally reach Earth's surface are relatively harmless. Most are absorbed, thereby heating the surface which, in turn, radiates photons to space. Whereas the incident photons, from the hot

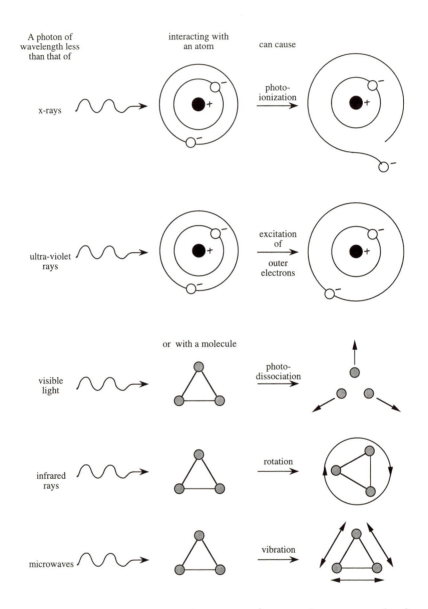

Figure 3.7 Possible interactions between a photon and atom or molecule. The most energetic photons (with the shortest wavelength) are at the top of the figure; toward the bottom, energy levels decrease, and wavelengths increase.

Sun, are very short and energetic, those from Earth's relatively cool surface are long, infrared photons with less energy than the incident ones. They generally do not ionize or photodissociate molecules, but they can cause atoms to vibrate more energetically or molecules to rotate more rapidly as in the lower panels of figure 3.7.

A molecule does not simply absorb each and every photon it encounters but is selective. It has to be, because it cannot rotate at any rate, or its atoms cannot vibrate at any rate, only at rates with certain discrete values; the possible states of a molecule or atom are quantized. The orbits of electrons around the nucleus of an atom are similarly quantized; only certain orbits are possible. It follows that a gas can absorb only the photons that have exactly the quanta of energy needed to change a molecule from one state to another, or to bump an electron from one orbit to another. If light with a continuous spectrum of colors—from a white hot body for example—passes through the gas sodium, only a few colors—yellow, for example—are absorbed. If it passes through potassium, red rather than yellow is absorbed. To revive our earlier analogy, sodium accepts only a few denominations, $5 and $1 bills, potassium only $1 bills and dimes. Some gases introduce numerous lines into a continuous spectrum; they accept several denominations: $100 bills to ionize them; $10 bills to photodissociate them; $5 bills to make them vibrate more energetically; $1 bills to make them rotate more rapidly; pennies to make them flap their wings more rapidly.

The more complex a molecule, the more moves (vibrational, rotational, etc.) it has, and the more readily it interacts with a variety of photons. Relatively simple diatomic molecules have a small repertoire. That is why nitrogen (N_2) and oxygen (O_2) can interact only with the big spenders in sunbeams, the energetic ultraviolet rays that photodissociate and ionize these gases high in the atmosphere. The more complex triatomic molecules such as carbon dioxide (CO_2), water vapor (H_2O), and ozone (O_3) have a larger repertoire. Ozone, for example, interacts with both short ultraviolet rays on their way to the Earth's surface and long infrared rays from Earth's surface. Although nitrogen and oxygen are by far the most abundant gases in our atmosphere at present, neither interacts with infrared earthlight on its way to space.

Greenhouse Gases

Earth's surface is at such a low temperature, generally less than 30°C, that its radiation is in the form of infrared heat. Those photons have so little energy that they are unable to interact with the diatomic molecules that account for most of the atmosphere, oxygen and nitrogen. The triatomic and other complex gases with which the infrared photons do interact—known as greenhouse gases—can be identified by inspecting the spectrum of radiation that Earth sends to space. Figure 3.8 shows this spectrum as measured by a satellite as it passed over the island Guam in the tropical Pacific Ocean. The jagged line is the observed spectrum of colors that Guam radiates to space. The smooth lines correspond to the radiation it would emit if its atmosphere had no greenhouse gases. Each smooth line corresponds to a different temperature. The warmer the surface, the more heat it radiates. Temperatures at the surface of Guam are close to 27°C, and part of the observed curve is indeed close to the smooth one that corresponds to a surface at that temperature. The departure from the smooth curve is most striking in a gaping hole centered on a wavelength near 15 microns. The gas carbon dioxide, which absorbs that particular wavelength may be present in modest amounts—it accounts for 0.035% of the atmosphere—but it clearly is an effective absorber of infrared radiation. From figure 3.8, we can infer that carbon dioxide absorbs primarily at elevations where the temperature is approximately −50°C.

Guam's radiation curve shows that there is also prominent absorption at wavelengths near 10 microns. The gas ozone absorbs those waves. Previously we encountered ozone as an absorber of short, energetic ultraviolet rays. Now we find that this voracious molecule also captures certain long infrared rays; it is a powerful greenhouse gas.

Different greenhouse gases make different contributions to global warming. The addition of another molecule of CFC to the atmosphere will contribute more to global warming than will another molecule of carbon dioxide. One reason for this is evident in Guam's radiation curve. The curve shows that there are windows through which photons with the right wavelengths can travel unimpeded through the atmosphere. One such window stretches from a wavelength of 10 microns (where ozone absorbs) to one of 15 microns (where carbon dioxide absorbs). The introduction of greenhouse gases that absorb colors between 10 and 15 microns will shut the window through which infrared heat escapes unimpeded at present. Shutting the window will enhance global warming considerably. Additional carbon dioxide or ozone molecules will also amplify the greenhouse effect but, because the atmosphere already has a fair amount of carbon dioxide and ozone, not as efficiently as CFC molecules that absorb in the open window. The CFCs are therefore doubly dangerous. They cause the

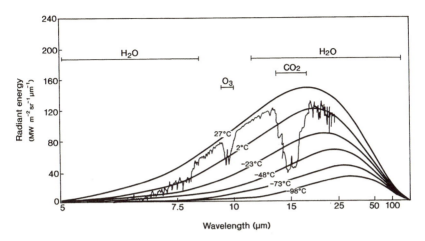

Figure 3.8 The jagged curve is Earth's radiation as measured by a satellite over the island Guam in the tropical Pacific Ocean, where the surface temperature is near 300°K (or 27°C). The smooth curves show the expected radiation from surfaces with the indicated temperatures. Much of the measured radiation is from the surface of Guam, except that for certain wavelengths, the radiation appears to be from regions with much lower temperatures, presumably from the upper atmosphere. The atmospheric greenhouse gases that absorb the surface radiation and, in turn, radiate at lower temperatures include: carbon dioxide, at wavelengths centered on 15 microns; water vapor over a broad spectrum of wavelengths; and ozone at wavelengths near 10 microns.

destruction of the ozone layer, and they are exceptionally efficient infrared absorbers that can contribute disproportionately to the greenhouse effect.

A greenhouse gas far more important than any identified thus far is the versatile chemical water, an efficient absorber of infrared radiation over a broad spectrum of colors. (The microwave ovens found in most kitchens bombard food with photons that vibrate the water molecules in the food so energetically that the water boils. That is why bread in microwave ovens becomes soggy.) To pinpoint the extent to which water vapor in the atmosphere contributes to greenhouse warming is equivalent to shooting at a moving target because the contribution depends on the temperature of the atmosphere. The warmer the atmosphere, the more water vapor it tends to have (assuming there is a readily available source of water such as the oceans). Thus, an increase in atmospheric temperatures is likely to increase the amount of water vapor in the atmosphere. Once there is more water vapor, temperatures increase even further. This escalating tit-for-tat can get out of hand rapidly. That happened on Venus but not on Earth, a topic explored in chapters 5 and 10.

The Thermal Structure of the Atmosphere

If the atmosphere were simply a blanket that is transparent to sunlight but opaque to infrared heat from the Earth's surface, then temperatures should be at a maximum at the surface and should decrease steadily with increasing height. In reality, the atmospheric temperature profile is much more complex. Measurements by means of balloons, rockets, and satellites reveal the strikingly curvaceous profile shown in figure 3.9. The undulations define layers, known as the troposphere, stratosphere, mesosphere, and thermosphere, in which temperatures alternately decrease and increase with height. To explain this profile we have to take into account a factor neglected thus far: the absorption of sunlight on its way to Earth's surface.

If all the rays in sunlight were to penetrate to Earth's surface, then we would find ourselves in a dangerous situation because some of those rays, specifically the ultraviolet ones, can seriously harm a variety of life-forms, including humans. Fortunately, our atmosphere filters out the dangerous rays before they reach the surface. The absorption of those rays from above generates heat and causes temperatures to increase with increasing elevation. (This contrasts with the greenhouse effect, which involves the absorption of infrared rays from below and hence temperatures that decrease with elevation.) In figure 3.9, we can therefore identify the thermosphere and stratosphere as the main atmospheric layers that absorb sunlight.

Sunlight has a spectrum of colors that ranges from short, dangerous ultraviolet rays, to the rainbow colors that constitute visible light, to relatively harmless infrared rays. The shortest waves pose a threat to life but their photons are so energetic that, at heights of 200 km and more in the thermosphere, they are absorbed by nitrogen and oxygen molecules that are photodissociated and ionized in the process. Therefore, at great altitudes, there is an ionosphere (formed of atoms that lack electrons), and a greater abundance of oxygen atoms (O) than oxygen molecules (O_2). The upper atmosphere may be too rarefied for human comfort—with elevation, the density of air decreases so rapidly that balloonists and mountaineers risk suffocation should they rise too high—but the gases there are nonetheless of vital importance to our well-being because they absorb ultraviolet rays. The thermosphere is literally our first line of defense.

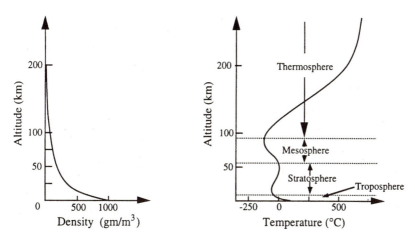

Figure 3.9   Vertical variations in the density and temperature of air as determined by unmanned balloons, rockets, and satellites. Temperature alternately decreases and increases to define layers known as the troposphere, stratosphere, mesosphere, and thermosphere. The density of air varies in a much simpler manner: it decreases rapidly with elevation throughout the atmosphere, but that is not necessarily true of the concentration of certain gases. Chemical reactions that produce ozone in the stratosphere cause its concentration to be at a maximum in that layer.

Ultraviolet rays that succeed in penetrating the thermosphere next encounter the stratosphere, our second line of defense. The reasons for the abundance of ozone in the stratosphere where it absorbs ultraviolet rays is deferred to chapter 12.

Aurora

The Sun radiates light, streams of photons, and, in addition, ejects streams of slower charged particles, primarily electrons and protons. They constitute the solar wind, which has speeds of 400 to 500 km per second. Light from the Sun reaches us within a matter of minutes; the solar wind takes a few days. The particles in the solar wind, though slow, are massive and are capable of doing more harm than energetic ultraviolet photons. Fortunately, Earth is protected by its magnetic field. This field is similar to that of a bar magnet located near Earth's center, but it is caused by the motion of molten iron in Earth's hot interior.

The interactions between Earth's magnetic field and the solar wind give the magnetic field lines the shape of a comet. The magnetic field deflects most of the dangerous particles in the solar wind away from Earth and guides the more determined ones toward Earth's polar regions. There they meet their nemesis, the atmosphere, in a fiery and spectacular display—curtains of light that unfold over thousands of kilometers to reveal greenish-white colors with splendid touches of pink or magenta. This is the aurora, which hangs high in the atmosphere at altitudes of 100 km and more and signals the interception of dangerous intruders by gas molecules that dart to and fro. The magenta comes from hits by nitrogen molecules, the green from hits by oxygen atoms.

Occasionally, the Sun has outbursts. It sends flares and an intensified solar wind in this direction causing the magnetic field on Earth to tremble erratically and creating problems for communication lines, radar systems, and long-distance power transmission lines. Fortunately the damage is limited because the atmosphere can withstand the dangerous assault; the aurora shines brighter than usual.

The geologic record indicates that Earth's magnetic field sometimes reverses polarity: the South Pole becomes the North Pole and vice versa. When this happens, there is a brief period when Earth has no magnetic field. On such occasions, the intruders can overwhelm the defenders and can contribute to the extinction of species.

The interactions between light and air discussed in this chapter can explain phenomena such as the greenhouse effect and the thermal structure of the atmosphere. The explanations, however, are only qualitative because they neglect atmospheric motion, which redistributes heat and matter vertically, thus cooling off Earth's surface. Chapters 4 and 5 concern this topic, convection, first in a dry atmosphere, then in one that has moisture and clouds.

4

WHY THE PEAK OF A MOUNTAIN IS COLD

AEDALUS AND ICARUS, father and son, tried to flee from Crete to Sicily on wings of feathers and wax. The son flew too close to the Sun, his wax melted, and he plunged to the ground.

This enchanting story, about a cautious old man, probably too wise to learn anything new, and an adventurous youth who should be applauded for being daring even though it cost him dearly, is at variance with the experience of those who rise through the atmosphere by climbing a mountain or, less strenuously, by ascending in a hot-air balloon. Figure 4.1 shows that the two Englishmen who chose the latter mode of transportation to rise to an altitude of nearly 9000 meters on June 26, 1863 found themselves in distress but, unlike Icarus, not because of excessive heat. One of them is seen trying to close a valve by pulling at a string with his teeth; his limbs are frozen stiff. The other is unconscious because of a lack of oxygen. (Both men survived the flight.)

The balloonists, as they rose higher and higher, approached the sun and therefore absorbed more and more heat. This would have caused their temperature to increase, had it not been for the loss of heat to the surrounding air. They were rising through an ocean of air, Earth's atmosphere, which is comfortably warm near Earth's surface but is frigid aloft. They lost so much heat to the cold air that surrounded them that it was of little consequence that they were slightly closer to the sun. By rising too high, they ran the risk of freezing to death. Furthermore, the density of air decreases so rapidly with altitude that they had difficulty breathing.

To explain what happened to the balloonists in the figure, we must address two questions: why are the atmospheric gases so confined to the immediate vicinity of Earth's surface that pressure decreases rapidly with elevation, and why do temperatures fall with an increase in height? A tentative answer to the latter question is available from the discussion of the greenhouse effect in chapter 3, which explains that the atmosphere is a blanket that keeps Earth's surface warm. Understandably, temperatures are higher in the lower part of the blanket than in its upper parts. Although this argument yields results that are

Figure 4.1 On June 26, 1863, two English balloonists rose to an altitude of 29,000 feet. One fainted from lack of oxygen. The other had to open a valve with his teeth because his arms were paralyzed by the cold. From Glaisher (1871).

qualitatively correct, the numerical values for the rate at which temperatures should decrease with height are in poor agreement with measurements. The calculated rate of decrease is far larger than what is observed. Furthermore, the average temperature calculated for the surface of the Earth, 67°C, is far higher than the actual average temperature of the surface which is 15°C. To keep the Earth's surface cool, heat must flow from its lower to its upper layers.

Heat can flow from a warm to a cold body in three ways. The arguments in chapter 3 involve strictly radiation, which is possible across a vacuum. (Radiation is how heat from the Sun reaches Earth.) Conduction requires physical contact between the warm and cold bodies because the most energetic molecules induce their immediate neighbors to be less lethargic. Convection occurs in fluids when a warm parcel consisting of a huge number of energetic molecules, moves through the fluid from one region to another. The latter process is particularly important for the vertical redistribution of heat in the atmosphere. Earth's surface, which is mostly water (oceans), loses a large amount of heat through evaporation, a process in which water at the surface is transformed into the gas water vapor just above the surface. Should the water vapor accumulate next to the surface, evaporation would come to a halt. Convection is of critical importance in transporting the water vapor to greater elevations in the atmosphere.

Convection is evident whenever cold water in a pot is put over a flame so that the highest temperatures are at the bottom. The water at the bottom of the pot, because it is warm and buoyant, rises spontaneously and is replaced by colder water from above. This motion redistributes heat throughout the pot so that the water ultimately has a uniform temperature. The atmosphere is similar to the water in the pot because both are heated from below. Solar radiation penetrates to Earth's surface, which it heats. Once the surface is sufficiently hot, the air above it starts to rise spontaneously.

In a pot of boiling water, convection results in a uniform temperature throughout the pot. In the atmosphere, on the other hand, convection results in temperatures that decrease with elevation because, as they rise, parcels of air expand as the air becomes more rarefied with height. (The density of air decreases with elevation because Earth's gravity traps most of the air molecules close to the surface of the Earth.) The expansion of a rising parcel of air requires the expenditure of energy, specifically thermal energy, so that the temperature of the rising, expanding parcel of air decreases. Hence, a parcel that starts at the warm base of a mountain is cool by the time it reaches the peak.

Convection is instrumental in the vertical mixing not only of heat but also of matter because it involves the vertical circulation of parcels of air composed of many different molecules. That is how dense CFC molecules succeed in rising to the stratosphere. Those who argue that CFCs molecules are too heavy to reach the stratosphere, and hence are unable to cause the ozone hole, are unfamiliar with convection. This process has a profound effect on atmospheric conditions, especially at Earth's surface, because it redistributes matter vertically. For example, the frequent absence of convection in certain regions, such as Los Angeles, causes pollutants to remain near the ground where man releases them. Conditions that inhibit the vertical mixing of air are known as *inversions* (of the normal state of affairs presumably). The smoke from a chimney often indicates the presence of convection: smoke rises and falls in huge loops when convection is present and drifts with the wind without dispersing when it is absent (fig. 4.2).

Figure 4.2   Plumes from a chimney when atmospheric conditions favor (left) and inhibit (right) convection.

The "Height" of the Atmosphere

The molecules of a gas are free to move about so that they usually fill whatever space is available. For example, the gas in a jar can expand to fill an entire room. This happens rapidly once the jar is opened because the molecules move extremely fast. An oxygen molecule in the air near Earth's surface can travel at speeds of 4000 miles per hour! It does not go far before it collides with another molecule because there are a great many of them, approximately a million billion molecules in a cubic foot of air at sea level. One molecule collides with another very often, nearly 100 billion times per second. Because of these random collisions they usually distribute themselves uniformly throughout the available space. In that case, why do balloonists and mountaineers find that the density of air decreases with height? What keeps air molecules close to Earth's surface?

When Newton noticed the apocryphal apple fall from the tree, his brilliant insight was not that gravity pulled the apple down—many people were probably aware of that—but that gravity reaches into space and attracts even the distant Moon toward Earth. Minuscule, fast-moving molecules high in the atmosphere are also pulled downward. It keeps the molecules relatively close to Earth's surface and causes their abundance to decrease with elevation. The rate at which their abundance decreases depends on a compromise between gravity, which pulls molecules downward toward the surface of the Earth, and the tendency of the molecules to move about freely. This compromise determines the effective height of the atmosphere. Because gravity keeps most of the air molecules close to the Earth's surface, the column of air above us, which weighs down on us, becomes lighter when we climb a mountain or rise in a balloon. In other words, both pressure and density decrease with elevation. (This would happen even if the gravitational attraction of Earth remained constant with increasing height above Earth's surface. In reality, gravity does decrease with distance from the surface of Earth but that factor does not contribute significantly to the small height of the atmosphere.)

A U-tube partially filled with a liquid, water or mercury, is a suitable instrument for measuring pressure. If both arms of the tube are open to the atmosphere, then, because the atmosphere weighs equally heavily on the two water surfaces, the water reaches exactly the same height in the two arms (fig. 4.3). If one arm is sealed after the air has been removed from it, then the water in that arm rises because nothing is pushing down on that water surface while the air continues to

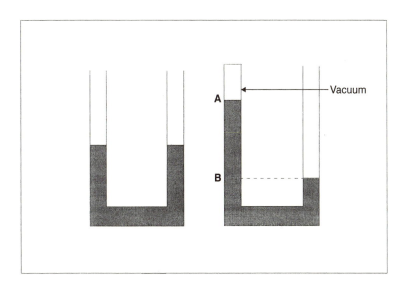

Figure 4.3   A simple barometer.

push down on the water surface in the other arm. The force with which the air pushes down on the water surface is equal to the weight of the column of water AB. At sea level, that column is 32 feet (more than 10 meters) high, which translates into a pressure of 1 kilogram per square centimeter. Air molecules that continually bombard the water surface exert that pressure. At the peak of a mountain, there are fewer air molecules that bombard the surface and the pressure is lower. (The barometer, the instrument that measures pressure, was invented in 1643 by Torricelli, a student of Galileo.)

In this discussion we interpret pressure in two different ways. First we regard pressure on a surface to be a consequence of the bombardment of that surface by gas molecules that move about randomly. Thus, the gas inside a balloon keeps its surface taut by exerting a pressure on it. If we squeeze the balloon, thus decreasing its volume and surface area, then the same number of gas molecules inside the balloon strikes the surface more often and hence exerts a greater pressure. Note that the density of the gas increases when the balloon is squeezed. Hence, an increase in the density of the gas is accompanied by an increase in pressure. Robert Boyle, a contemporary of Isaac Newton, first quantified this relation between pressure and density. Boyle's Law, is valid provided the temperature of the gas remains constant and is independent of any gravitational force. If we squeeze a balloon in a weightless space ship, both the pressure and density of the gas in the balloon will increase.

If we happen to be in a gravitational field—of Earth, for example— there is a second way to interpret pressure: the pressure exerted on us by the air molecules that continually bombard us can be regarded as the weight of the column of air above us. This is analogous to the pressure exerted on a submarine by the weight of the column of water above it. As the submarine rises toward the ocean surface, the pressure decreases as the column of water above the submarine shrinks. The weight of a column of water depends on the height of the column and on the density of the water. The density is, for practical purposes, a constant so that pressure decreases as rapidly as does the height of the column above the submarine, as shown in figure 4.4.

If we take a barometer with us as we rise through the atmosphere in a balloon, we will find that the pressure decreases with elevation at a rate that exceeds that shown in the left panel of figure 4.4 (for a submarine rising to the ocean surface). This happens because, whereas the density of water is essentially constant, the density of air decreases with elevation. Hence, as we rise in the atmosphere, the weight of the column of air above us decreases, not only because the column shrinks, but also because the air becomes less dense.

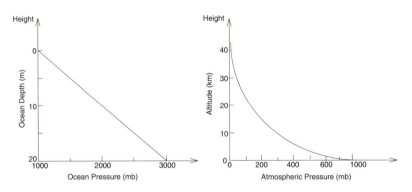

Figure 4.4   *Left,* In the ocean, which has an essentially constant density, pressure increases linearly with depth. *Right,* In the atmosphere, both pressure and density decrease exponentially with elevation.

The gases in the atmosphere are subject to two laws. One is Boyle's law: pressure must remain proportional to density (if the temperature does not change). The other is the barometric law, which states that the pressure at a point depends on the weight of the column of air above that point. Boyle's law would permit the gases of the atmosphere to expand indefinitely (provided their density and pressure decrease at the same rate), but they are subject to another constraint, the effect of gravity as expressed by the barometric law. These constraints can be expressed as mathematical equations whose solution describes the compromise that is reached between the opposing tendencies of air molecules, to expand upward and to remain close to Earth's surface. That compromise determines the depth of the atmosphere. The result is shown in the right panel of figure 4.4.

Figure 4.4 shows a clear difference between the manner in which pressure decreases with elevation in water and in air. The origin of this difference is the different degrees of freedom that air and water molecules have. The competition between gravity, which pulls molecules downward, and the tendency for molecules to move randomly is an unequal one in the case of water. Whereas the molecules of a gas are rebellious and constantly try to move upward, those of a liquid are subservient and remain below a horizontal surface, which is why the ocean has a definite upper surface but the atmosphere does not. The rate at which density decreases with height in the atmosphere depends on the local value of the density and becomes smaller and smaller with elevation so that density never reaches zero. The density is said to decay exponentially. Pressure is proportional to density and behaves in the same manner. Air molecules become increasingly rare with increasing height, but at any elevation it is always possible to find a few molecules. Although the atmosphere has no top, no fixed height, it is useful to decide on an effective height in order to know at what altitude balloonists or mountaineers are likely to run out of air. One measure, which turns out to be reasonable for practical purposes, is the height of an atmosphere of constant density that exerts the same pressure as our atmosphere at Earth's surface. This height, which is known as the scale height of the atmosphere, is 10 km approximately. Over that vertical distance density and pressure, in reality, decrease by 37% approximately. If pressure at the surface is 1 atmosphere, then it is 0.37 atmospheres at a height of 10 km, 0.14 (= .37 × .37) at 20 km, 0.05 (= .14 × .37) at 30 km, and so on. The decrease in density with altitude is so rapid that almost 80% of the mass of the atmosphere and practically all the water vapor and clouds are within 10 km of Earth's surface, a distance less than 0.002 Earth radii. This explains why, from space, Earth's atmosphere appears as a very thin veil of gases that envelopes the planet. The air becomes so rarefied at heights near and greater than 10 km where balloonists and mountaineers on Everest run into trouble—that commercial jets, which fly at those heights, are pressurized and have oxygen masks available for emergencies should a cabin puncture so that air from inside rushes out.

We expect the scale height of the atmosphere to increase as the temperature increases because the higher the temperature, the more energetically the molecules move about. If, on the other hand, we can arrange for the force of gravity to increase—by moving to a more massive planet, for example—then we expect the scale height to decrease. This dependence of scale height H on temperature T and gravity g is expressed succinctly by the following formula, which also shows the dependence on the molecular weight m of the gases in the atmosphere.

$$H = \frac{RT}{mg}.$$

The formula indicates that H is smaller for a heavy gas such as carbon dioxide, which has a large value of m, than for a light gas such as hydrogen. In other words, gravity traps heavy carbon dioxide molecules closer to the ground than it traps light hydrogen molecules. (For the mixture of gases in the atmosphere, H has a value of approximately 10 km, but H has different values for individual gases.) From this result, some people conclude that very heavy molecules such as CFCs are so strongly trapped near Earth's surface that they are absent from the upper atmosphere and cannot harm the ozone layer. To ap-

preciate why this inference is incorrect, we have to examine the assumptions we made in deriving the expression for H, especially the
one that the atmosphere has a uniform temperature. We next relax
that assumption.

Why Temperatures Decrease with Height

Heat usually flows from a warm to a cold object. Why then does it
not flow from the lower part of the atmosphere to the elevated parts?
It can do so in three different ways; radiation, conduction, and convection. Radiation is possible across empty space, from the Sun to
Earth, for example. Conduction requires physical contact between the
objects that exchange heat. Convection is possible only in fluids and
gases in which the flow of heat from one region to another involves
the movement of matter from one region to the other. The atmosphere
is a gas and can avail itself of all three modes of heat transport. Why
is the upper atmosphere nonetheless cold?

The atmosphere is heated from below because the Sun's rays penetrate through the atmosphere and heat Earth's surface. This is in striking contrast to the oceans, which absorb the rays of the Sun close to
the ocean surface so that the ocean floor remains pitch dark. The different ways in which the Sun heats the atmosphere and the ocean are
similar to the two ways in which a pot of water can be heated: the pot
can be either over the flame or under the flame.

If the flame is over the pot, then the water molecules near the surface vibrate more energetically and, in due course, induce their neighbors at slightly greater depths to move with more abandon. Heat
gradually diffuses downward, toward the bottom of the pot. Although heat flows downward, there is no net motion in that direction.
There are only molecules that move randomly without going anywhere. This mode of heat transfer, conduction, depends on physical
contact between molecules. In a gas, conduction depends on the frequency with which molecules collide. If the density of a gas is low,
conduction is particularly inefficient because collisions between molecules are infrequent.

Heating from above, by means of conduction, is less efficient than
heating from below. When water is boiled by putting it in a pot over a
flame, the molecules at the bottom of the pot are the first ones to start
vibrating more energetically. This causes a "parcel" that consists of a
huge number of these molecules to expand somewhat so that it becomes less dense than the colder parcels above it. If it is less dense
than its surroundings, the parcel is buoyant and starts to rise. This
means that there are now two types of motion: disorganized motion

of molecules, and the organized, upward motion of ensembles or parcels of molecules. (Imagine children running back and forth in a train that moves forward in an orderly manner.) The organized motion, known as convection, distributes warm parcels of water throughout the pot and ensures that new parcels of cold water come into contact with the hot bottom of the pot. It is a very efficient way of transferring heat throughout the container and is the reason why heaters in a cold room should be near the floor: the heated air rises rapidly. Air conditioners should be near the ceiling: cooled, dense air sinks readily. In the atmosphere, convection performs the same function. It redistributes heat vertically, taking it from the surface to higher elevations, and also redistributes air molecules. Gravity traps most air molecules near Earth's surface and, in the absence of organized atmospheric motion, would be most successful with molecules of dense gases. (In the absence of organized motion, the heavy molecules of carbon dioxide and CFCs would be trapped much closer to the surface than those of light hydrogen, for example). The organized motion of convection transports parcels of air, parcels that include heavy and light molecules, to considerable elevations. That is how heavy molecules can reach great altitudes. In a thunderstorm that has intense convection, parcels of air can rise from Earth's surface to heights in excess of 10 km in a matter of minutes. (The vertical motion can be so intense that it poses a danger to airplanes.) Because of gravity, the density of air always decreases with height; because of convection, a sample of air from aloft has essentially the same composition as a sample from near the surface.

Conduction is subtle: heat flows in a certain direction without net movement in that direction. Convection is rebellious: it destroys the very conditions that favor it. When a flask of water is warm at the bottom and cold at the top, convection arises spontaneously and redistributes heat until the water is at a uniform temperature. At that stage, convection ceases. It is appropriate that scientists describe the conditions that favor convection as *unstable*. The test for instability is simple: a parcel of water is displaced upward to check whether it is more or less dense than its surroundings. If it is more dense, it sinks back and conditions are stable. If it is as dense as its new surroundings, and happily remains there, then conditions are neutral. If it is less dense than its surroundings, then it is buoyant, continues to rise, and drifts far from its origins. Such conditions are unstable and lead to convection. The density of water depends on its temperature in a simple manner and, instead of checking how density varies with height, it is possible to check how temperature varies: if warmer water is below colder water, convection occurs. Neither water at a uniform temperature of 80°C nor water at a uniform temperature of 20°C is a likely environment for convection. What matters is not the absolute temperature but temperature differences within the water. In a graph, the slope of isotherms matters, as in figure 4.5. Rebellious convection can be suppressed by making sure that a fluid is stably stratified. A parcel that is elevated then falls back to its original position. Vertical mobility is inhibited, and motion is confined to horizontal layers.

The atmosphere is heated from below, the way water in a pot on top of a flame is heated. Why then does convection not distribute heat vertically in the atmosphere so that its temperature is uniform? Why is the peak of a mountain colder than its base? For an answer, we must start with the test for convection. If a parcel of air were elevated slightly, would it be less dense than its new surroundings? In the case of water, a parcel that is elevated barely changes its volume—water is practically incompressible—so that the density of the parcel does not change. Hence we need only compare the density of water at the original and new positions. In the case of air, the test is not as straightforward because the parcel of air, when elevated, expands and cools.

The parcel expands when elevated because pressure decreases with height. It is common experience that when air expands by escaping from a bicycle tire it cools off. When it is compressed, by pumping it into the tire, temperatures rise. To explain why, we must explore the

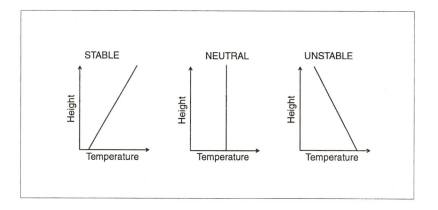

Figure 4.5 In a stable situation, the coldest, and hence densest, water is at the bottom, the warmest and least dense water is at the top. If this state of affairs is reversed, the situation is unstable and will rapidly change to one of neutral stability in which temperature and density are uniform, as in the middle panel.

nature of energy which comes in various forms: kinetic, which is associated with motion; potential, which is associated with position; and a few other forms that are discussed below. One form of energy can be converted into others, but such transformations leave the net amount of energy unchanged. It is remarkable that nature chooses to maintain a constant value for something as abstract as energy, but this fact is of enormous practical importance.

The simplest tool that depends on the conservation of energy is a lever; it involves only gravitational energy. With a lever a baby can lift an elephant by exchanging gravitational energy. The baby of mass m has gravitational energy mgh by virtue of being at a height h above the ground. (Gravity g must enter the expression for gravitational energy because, in a weightless environment, nobody has that type of energy.) The baby can raise an elephant of mass M to a height H by spending its gravitational energy, which the elephant gains. Hence,

$$gMH = gmh.$$

This application of the conservation of energy is so simple that it was discovered long before scientists introduced the concept of energy. The ancient Greeks knew of the usefulness of levers, as did Leonardo da Vinci, who rediscovered these ideas. Next consider the pendulum, which involves two forms of energy, gravitational and kinetic. When the pendulum starts its downward swing, it is motionless, has no kinetic energy, and has maximum gravitational energy. At the lowest point of its arc, it has lost gravitational energy but is moving fast and has considerable kinetic energy. As the pendulum swings back and forth, there is a continual change from gravitational energy, to kinetic energy, back to gravitational energy, but something also remains constant. The total energy, the sum of kinetic and gravitational energy, remains constant. From the conservation of total energy, it is possible to calculate the speed of the pendulum at different positions with great accuracy.

In the nineteenth century, scientists discovered yet another form of energy, heat, which can be converted into and can be derived from other forms of energy. In one of the experiments of the British physicist James Joule, the gravitational energy lost by a weight that falls a certain distance was used to drive a paddle that stirs water, causing its temperature to increase. This demonstration led to acceptance of the first law of thermodynamics, which states that heat is a form of energy and that its conversion into other forms of energy is such that total energy is conserved. For example, a girl who jumps from a diving platform into a swimming pool first converts her potential energy into kinetic energy and then converts that kinetic energy into thermal

energy of the water in the pool. By using the first law of thermodynamics, it is possible to calculate by how much, the temperature of
the water increases.

When applying the first law of thermodynamics to explain certain
phenomena, it is of extreme importance to keep in mind the relation
between energy and work. To a scientist, work always involves the
use of force to move an object, whether it be lifting a book, or pushing a car, or driving a nail into wood. The last example illustrates how
work involves a change in energy; a hammer loses kinetic energy to
do work on the nail. A parcel of air that expands and pushes its
surroundings back loses energy in the process. That energy comes
from the random motion of the air molecules in the parcel. After the
expansion, the molecules move less energetically. In other words, the
expansion causes the temperature of the air to decrease. Conversely,
compression of a parcel of air leads to higher temperatures.

We are finally in a position to explain why the peak of a mountain
is colder than its base. Recall that the atmosphere is heated from below, like a pot of water over a flame. Heating from below, whether it
be of water or air, brings into play rebellious convection, which redistributes heat until there is a state of neutral stability. During convection, warm parcels rise and take their thermal energy with them, thus
bringing heat to higher elevations. In the case of water, which is essentially incompressible, the parcels change neither their volume nor
temperature as they rise so that the final result of convection is water
at a uniform temperature. In the case of a compressible fluid such as
air, parcels that rise expand because of the decrease in pressure. This
expansion uses up some of the thermal energy of the parcel so that its
temperature falls. Hence, if convection circulates air throughout the
atmosphere, then the final result is temperature that lapses with
height. That is why the peak of a mountain is cold.

The lower the temperature, the less energetic are the random movements of the molecules. Does that imply that there is a temperature at
which the molecules do not move at all, at which everything is dead?
To determine such a temperature, we first need an instrument that
measures temperature, a thermometer. The most common thermome-
ters exploit the tendency of certain materials to expand when heated.
Often the material is mercury in a narrow glass tube with marks that
indicate different temperatures. The mark for 0°C indicates the length
of the column of mercury when the tube is immersed in ice water.
The mark for 100°C is obtained by immersing the tube in boiling water. Interpolation between these two reference temperatures yields a
scale for the thermometer provided we assume that mercury expands
uniformly when it is heated. Not all liquids have this property. Water,

for example, expands when heated above 4°C but it also expands when cooled below 4°C! This means that water has a maximum density near 4°C. Ice at 0°C is less dense than liquid water and therefore floats on water. This is fortunate for fish in a frozen lake—they survive in the water below the ice—but it can sometimes have fatal consequences, for the Titanic for example.

Water is an inappropriate material for use in a thermometer because it expands nonuniformly. In principle the best material, but an impractical one, is a gas because its volume expands remarkably uniformly with temperature provided the pressure remains constant. On a graph, this relation between volume and temperature corresponds to the straight line between the small circles in figure 4.6. The small circles show the volume of the gas at 100°C, the boiling point of water, and at 0°C, the freezing point of water. The line indicates that, as the gas cools, its volume decreases, provided its pressure remains constant. The French chemist and inventor Jacques Charles, made this discovery in 1787. (Charles also repeated, with his own innovations, Benjamin Franklin's experiments with lightning. Franklin, while in France, visited Charles and congratulated him on his work.)

The graph makes it easy to extrapolate Charles' results to very low temperatures and to deduce that, at a temperature of −273°C, the volume of a gas vanishes! (The accuracy of this value can be questioned on the grounds that, before its volume disappears, the gas would have become first a fluid, then a solid.) In the molecular theory of matter, in which the temperature of an object measures how energetically the molecules move and vibrate, the random motion slows as temperatures decrease and almost ceases at −273°C, a temperature of maximum stillness. Theoretically, temperatures below absolute zero are impossible so that it is useful to add the number 273° to all Celsius temperatures, thus creating a new temperature scale, the Kelvin scale. Scientists have reached temperatures close to absolute zero and have discovered that very cold materials are capable of strange behavior. For example, some materials lose all resistance to electrical currents and become superconductors.

Charles discovered that a gas expands uniformly as its temperature increases, provided its pressure remains constant. The discussion of why the atmosphere is confined to the neighborhood of Earth's surface mentioned Boyle's law, which relates the pressure and density of a gas, provided its temperature remains constant. The marriage of Charles' and Boyle's laws provides a law that governs the behavior of an ideal gas; it tells us how the pressure, density, and temperature vary simultaneously.

We now have all the information needed to calculate the rate at which temperature decreases or lapses with height, namely the three

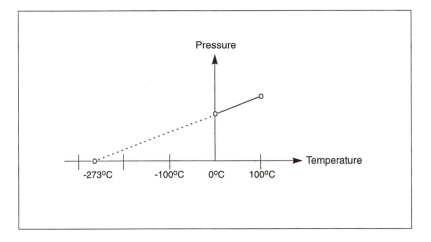

Fig 4.6 The dependence of the pressure of a gas, which is contained in a given volume, on temperature.

laws that relate density, pressure, and temperature: the ideal gas law, the barometric law, and the first law of thermodynamics. When expressed mathematically, they determine how these fields vary with height. The results indicate that temperature decreases linearly with height, at a rate known as the *adiabatic lapse rate*. (The term *adiabatic* indicates that no heat is added to or subtracted from a parcel as it rises, expands, and cools.) Its value is approximately 10°C per kilometer. All else being equal, Denver, the city that is a mile high, should be approximately 14°C cooler than New York or San Francisco at sea level.

The average pressure and density of air decrease with elevation in a manner that is not strongly affected by convection. However, convection does influence the manner in which the densities of different gases in air decrease with height, because it involves the vertical circulation of parcels of air that include light and heavy molecules. Convection is the means by which heavy molecules reach considerable altitudes. Those who deny that CFCs can rise to sufficiently high altitudes to cause the ozone hole base their arguments on results that are valid for a static atmosphere in which there is no convection.

In summary, the explanation for the cold peak of a mountain hinges on four crucial points.

1. Sunlight penetrates to Earth's surface, which it heats. Because the atmosphere is heated from below, convection comes into play. The oceans, by contrast, are opaque to sunlight and are heated from above.

2. Convective heating involves the circulation of parcels of air. The parcels, each composed of a huge number of molecules which all vibrate and move randomly, rise from Earth's surface.

3. Elevation causes a parcel to expand. This occurs because gravity keeps most of the air molecules near Earth's surface so that pressure decreases with altitude.

4. Expansion requires work, which costs energy. The source of energy is the random motion of the molecules. In other words, the molecules move less energetically after expansion so that the parcel of molecules loses heat and its temperature falls. The higher a parcel rises, the more it expands and cools, even though there is no exchange of heat with the surroundings. The rate at which its temperature decreases with height is the adiabatic lapse rate.

The Second Law of Thermodynamics

If energy is conserved, then sooner or later the Sun will stop shining. It cannot create energy from nothing. How much longer will it continue to shine, and for how long has it been shining? In the nineteenth century scientists addressed these questions because of a debate about the age of Earth. Some geologists, members of the Uniformitarian school, argued that Earth in the past was as it is today and refused to consider whether Earth or the Sun had an origin. It is an important principle in geology that the processes that have changed the surface of Earth in the past are still doing so today, but the originators of this idea went too far. The Sun could not have shone forever, and Earth could therefore not have existed forever. The Scottish physicist William Thompson (also known as Lord Kelvin of the Kelvin temperature scale) estimated that the Sun could have been producing energy at the current rate for approximately 100 million years so that that must be the age of the earth. In his calculations, Kelvin took into account the Sun's various sources of energy except for nuclear energy, which was not discovered until the twentieth century. The stupendous source that maintains the Sun's vast outpouring of energy is the nuclear fusion of hydrogen into slightly lighter helium. The Sun radiates light and loses mass in the process. (Einstein expressed the relation between mass m and energy E with his famous equation $E = mc^2$, where c is the speed of light.) Nuclear fusion has allowed the Sun to shine for close to 10 billion years and will allow it to shine for much longer.

The Sun supplies us with an enormous amount of energy and will continue to do so for a considerable time. Not all of it is available to us, however. To understand why, consider a speeding bullet. It has both kinetic energy, associated with the organized motion of the bullet as a whole, and thermal energy, associated with the disorganized motion of its atoms. The first can be converted into more of the second should a stone wall suddenly bring the bullet to a complete halt. After the collision, the kinetic energy is gone but the random vibrations of the molecules are more energetic than before so that the temperature of the bullet increases. Sometimes the rise in temperature is so high that the metal of the bullet melts.

The first law of thermodynamics, which enables us to calculate by how much the temperature of the bullet increases if it comes to a halt, is a powerful law, but it is also curiously lax. It allows any form of energy to be converted into any other form, as long as total energy is conserved. That provision sounds restrictive but can be permissive. It does not preclude the far-fetched possibility that the bullet can suddenly jump out from the wall, and speed backward, by converting the increase in its thermal energy back into kinetic energy. Kinetic energy is associated with orderly motion, thermal energy with disorderly motion. Order readily degenerates into disorder, but the reverse is not true. Heat is a degraded form of energy and, although it is possible to convert other forms of energy entirely into heat, it is impossible after that conversion to recover the original forms of energy in their entirety. Whereas the first law of thermodynamics states that no energy is lost when it is transformed form one form to another, the second law of thermodynamics recognizes that once energy is converted into heat, it is partially unavailable. Entropy, a concept introduced by nineteenth-century scientists, is a measure of the amount of energy that becomes unavailable when there is a conversion of energy from one form to another. The unavailable energy is associated with random fluctuation so that entropy is also a measure of chaos and disorder. The second law of thermodynamics states that, in a closed system, isolated from its surroundings, entropy either remains constant or increases.

If our universe is a closed system, then its entropy is increasing steadily so that more and more heat becomes unavailable until the universe dies a cold death. A system that behaves in the opposite manner, that becomes more and more orderly, so that its entropy decreases, cannot be a closed system and requires a source of low entropy. A beautiful example is the living body, an exquisitely organized structure whose maintenance and growth, which reduces disorder, requires low-entropy food. At the base of the food chain are plants, which grow by means of photosynthesis; they absorb low-entropy photons from the Sun, our ultimate source of energy and low entropy. None of the energy that Earth receives from the Sun is lost; the first law of thermodynamics insists that, to maintain an equilibrium, Earth radiate as much energy as it receives. However, there is a difference between the radiant energy that the Earth receives from the Sun and that which it emits to space. It receives energetic photons associated with short electromagnetic waves and radiates a larger number of less energetic photons associated with long infrared waves. Energy is conserved, but some of it becomes unavailable so that the exchange causes entropy to increase.

Atmospheric Pollution

Atmospheric convection, which redistributes heat and matter vertically, thus lowering the concentration of pollutants near the ground, does not occur continually throughout the day. It depends on high temperatures at the surface of Earth and hence occurs preferentially in the afternoons of sunny days.

Earth's surface gains heat from the Sun and loses heat because it radiates heat upward. After sunset it continues to radiate and therefore cools off. Because of conduction, the air in physical contact with the ground also cools so that the temperature profile changes, as shown in figure 4.7, within a kilometer or two of Earth's surface.

Toward the end of a sunny afternoon, there is warm air near the surface, cold air aloft. Early in the morning, when the reverse is true, conditions are far more stable because there is often cold air beneath warmer air. Such a state of affairs, known as an inversion, inhibits convection and can cause severe pollution from morning rush-hour traffic.

The lower atmosphere can have a complicated temperature profile because the winds, which vary with height, transport air masses with different temperatures from one region to another. The temperature profile determines the behavior of plumes from a smokestack. Consider a stable profile that requires temperatures to increase with height or to decrease slowly, at a rate less than the adiabatic lapse rate. Such a profile inhibits the vertical movements of air parcels so that the smoke from a stack does not disperse. If, on the other hand, the profile is unstable so that the air aloft is far cooler than that below, then the smoke should show evidence of convection, of energetic vertical motion. Temperature variations in a lake can similarly be inferred from the behavior of the effluent from a pipe below the surface of a lake. What matters in the lake is whether the temperature increases or decreases with height. Figure 4.8 illustrates various possibilities. It is assumed that the wind (or the current in the lake) moves from left to right. The effluent from the pipe (or smokestack) is as-

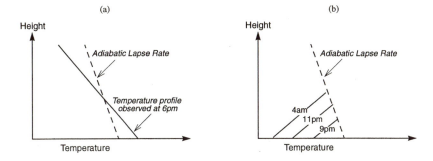

Figure 4.7 (a) By the end of a sunny day, the ground can have such a high temperature that the lower atmosphere has an unstable temperature profile. Convection then redistributes heat vertically to establish the profile that corresponds to the dashed line. (b) Subsequently, the ground radiates heat to space and cools off rapidly so that the temperature profile changes. Early in the morning, the lower atmosphere has a very stable temperature profile, a condition that favors the accumulation of pollutants near the ground.

sumed to have a density equal to that of the water (or air) at the top of the pipe.

Balloonists can experience conditions that range from the heat of summer to the cold of winter, from fog to rain to snow, by rising through the atmosphere without moving any significant distance horizontally. Such variations characterize the troposphere (changing sphere), the turbulent lower layer of the atmosphere that extends to a height of approximately 10 km. In this layer, temperature generally decreases with height; at greater elevations, as balloonists discovered early in the twentieth century, temperatures increase with height in a layer known as the stratosphere. The stable stratosphere is, in effect, an inversion layer over the troposphere and creates conditions analogous to those that lead to fumigation. Tall convective clouds can reach to the top of the troposphere but no higher. Unlike the troposphere, the stratosphere has essentially no turbulent weather, practically no clouds, and hence has perennial sunshine. Commercial planes fly just above the troposphere in the lower stratosphere in part to avoid the turbulence of the troposphere, thus ensuring smooth rides. The stability of the stratosphere means that if pollutants get there, they can remain for a long time. In the troposphere, where convection readily disperses pollutants, rain acts as a cleanser and washes out many of the pollutants, but that does not happen in the stratosphere.

This discussion of convection disregarded the presence of moisture in the atmosphere. The next chapter describes the profound way that this remarkable chemical influences atmospheric conditions.

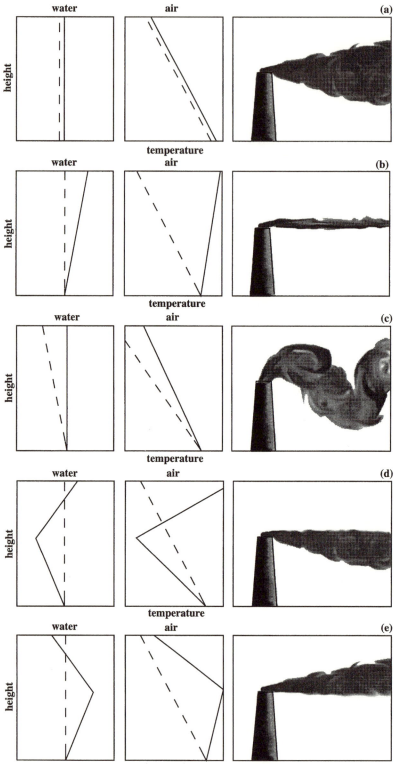

Opposite page:

Figure 4.8a *Coning,* a result of dispersion caused by the random motion of molecules, occurs under neutrally stable conditions. Such conditions prevail in a lake when temperature does not vary with depth. In the case of the atmosphere, temperatures must decrease with elevation at the adiabatic lapse rate. The dashed lines (in this and subsequent figures) correspond to neutral conditions. The solid line is the actual temperature of the atmosphere or water. The flow of air (or water) is assumed to be from left to right.

Figure 4.8b *Fanning* occurs when the stratification is stable, when there is said to be an inversion. A fluid parcel that is elevated falls back to its original position. Vertical but not horizontal dispersion is inhibited so that pollutants spread out in sheets in the direction of the wind or current. In the atmosphere, this often occurs during the early hours of the morning.

Figure 4.8c *Looping* is associated with unstable conditions that favor convection. The air that rises in some regions must necessarily be sinking in other regions so that the smoke rises and falls as shown.

Figure 4.8d *Fumigation* occurs when there are stable conditions (an inversion) aloft, unstable conditions below. Because of the stable conditions, upward dispersion is inhibited. The unstable conditions near the ground promote downward dispersion. On such occasions, atmospheric pollution is severe because pollutants are trapped in the lower layers. The inversion is essentially a lid on the convection below.

Figure 4.8e *Lofting* is a locally felicitous state of affairs because convection carries pollutants upward but not downward, which is why an increase in the height of smoke stacks can improve pollution problems locally (but exacerbate them downwind).

5

CAPRICIOUS CLOUDS

I am the daughter of earth and water,
And the nursling of the sky;
I pass through the pores of the ocean
    and shores,
I change, but I cannot die.
For after the rain when with never a
    stain
The pavilion of Heaven is bare,
And the winds and sunbeams with
    their convex gleams
Build up the blue dome of air,
I silently laugh at my own cenotaph,
And out of the caverns of rain,
Like a child from the womb, like a ghost
    from the tomb,
I arise and unbuild it again.
      (Percy Bysshe Shelley, "The Cloud")

CLOUDS have a great many names to match their infinity of shapes, but all are variations on three basic themes. Stratus clouds spread out horizontally to cover acre upon acre, appearing as an endless, sunlit prairie when seen from above. Cumulus clouds can grow vertically into tall towers that threaten with thunder and lightning. Insubstantial wispy cirrus clouds are transparent wind-blown webs of minute ice crystals at considerable altitudes, where they form a halo around the moon by passing in front of it. Most clouds are combinations of these three shapes and adopt Latin qualifiers—alto for high, nimbus for rain, and so on—to become cirrostratus, cumulonimbus, stratocumulus, etc.

Clouds enliven the sky by behaving capriciously, but in reality they are of crucial importance to the climate of our planet, sometimes heating it, sometimes cooling it. Clouds that cover huge areas reflect large amounts of sunlight, thus depriving Earth of heat. They also cause warming because water, in both liquid and gaseous form, absorbs infrared radiation from Earth's surface. Furthermore, clouds affect atmospheric temperatures by simply coming into existence or by disappearing.

Because they are composed of water, clouds can appear, disappear, reappear, and readily change shape. Whenever they arise "like a ghost from the tomb," the invisible gas water vapor condenses into visible water droplets or ice crystals. When clouds disappear, the reverse happens and water reverts to its gaseous form. Many substances are capable of similar transformations, changing their phase from solid to liquid to gas as their temperature increases. Water is exceptional because it is the only substance that is present in all three phases in the relatively narrow range of temperatures encountered near Earth's surface. Another unusual property of water is the large amount of heat that is involved when its phase changes. To transform a gram of ice at 0°C into water at 0°C requires 80 calories. That heat loosens the ties between the molecules so that rigid ice becomes a fluid in which molecules readily flow past each other. The further addition of 1 calorie for every gram of water raises the temperature by 1°C . Such heating causes the water to boil at 100°C, at sea level. To convert the gram of liquid water at 100°C into the gas water vapor at the same temperature requires 540 more calories. That considerable amount of heat is needed to break the ties between the molecules completely so that they are free as air. The cost of freedom is high, but the heat is latent and can be recovered when the molecules return to bondage. The latent heat that is required for evaporation, to convert a liquid into a gas, becomes available when the molecules of the gas reestablish their bonds and return to a liquid state. Thus, evaporation is associated with cooling, condensation with heating.

A wet swimmer on a windy beach feels cold when the water droplets on her skin evaporate. The ocean is similarly cooled when the winds evaporate water from its surface. That lost heat reappears when the winds carry the invisible water vapor upward until they condense into droplets of water, forming clouds. The appearance of clouds in the sky is accompanied by a release of latent heat, heat that originally came from the ocean. The ocean, in turn, gains its heat from the sun.

Earth's Energy Budget

The principal source of energy on Earth is light from the Sun. (It is far more important than the tiny amount of heat that reaches Earth's surface from its hot interior. Spectacular volcanic eruptions are too rare to contribute much heat globally.) Of the sunlight that Earth receives, approximately 30% is reflected back to space before it can heat the surface. Clouds are responsible for much of this enormous loss because they are white and cover vast areas. Snow and ice of glaciers also contribute significantly to the loss of heat.

In addition to reflecting so much sunlight, clouds, along with a few gases such as ozone, absorb some of the beams in sunlight so that only 50% of the sunlight that is incident at the top of the atmosphere is absorbed at Earth's surface. In a state of equilibrium, the surface loses as much heat as it gains. The losses occur in several ways, including the radiation of infrared rays by the surface. Much of that radiation is absorbed by the greenhouse gases in the atmosphere and is radiated back to Earth's surface. Clouds, of course, contribute enormously to the greenhouse effect by absorbing infrared radiation over a broad spectrum of wavelengths. Another way for the surface to cool off is by means of conduction, the transfer of heat to the lower layers of the atmosphere that are in physical contact with the surface. This transfer of *sensible heat*, as it is known, is particularly large in winter in regions where very cold air moves from a continent onto a much warmer ocean. Examples include air that moves from Asia onto the Pacific or from North America onto the Atlantic Ocean.

Earth's surface also loses heat when clouds carry heat from the surface to the upper atmosphere. This process starts when invisible water vapor evaporates from the ocean into the atmosphere, thereby cooling the ocean. Still invisible, the water vapor travels upward, by means of the winds, taking the latent heat with it. Aloft, the water vapor condenses into clouds and releases the heat originally acquired from the ocean. By means of this process, 1 m of water, on the average, evaporates from Earth's surface each year. (The globally averaged precipitation is 1 m and in a state of equilibrium evaporation must equal precipitation.) The evaporation of all this water causes Earth's surface to lose 83 watts per square meter, the equivalent of almost half the sunlight that reaches the surface. To appreciate the enormous importance of this loss, consider what would happen in its absence. The temperature of the surface would have to increase until the heat that is radiated upward equals that received from the Sun. In that case, the temperature, averaged over the globe, would be 67°C instead of the actual 15°C (fig. 5.1). Because clouds carry heat upward, the decrease in temperature with height is not as large as it otherwise would be.

The Hydrological Cycle

Clouds are capable of continually carrying heat from the surface to the air aloft, because they "change but . . . cannot die." Scientists pro-

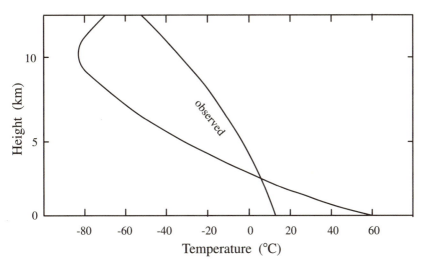

Figure 5.1   Earth's observed temperature profile, and the one it would have if the atmosphere were motionless and if its temperature depended on radiative processes only. In reality, the motion of the atmosphere effects a vertical transport of heat, removing it from the surface and depositing it aloft. As a result, globally averaged surface temperatures are a pleasant 15°C rather than an intolerable 67°C.

saically refer to this as the hydrological cycle, which begins when water evaporates from the oceans (fig. 5.2). Evaporation occurs primarily where sea surface temperatures are high and the winds are strong. Over land, evaporation is more complicated and depends on how well soil retains moisture from the most recent rains. The slope of a terrain and the type of soil—whether it is sand or clay, for example—are therefore critical factors as is the vegetation, which also transfers moisture from the soil to the air. The invisible water vapor, once it is in the atmosphere, drifts with the winds until condensation creates visible clouds. Clouds disappear because of evaporation or because water droplets coalesce until they are large enough to fall as rain, hail, sleet, or snow. Precipitation builds glaciers, feeds rivers, or sinks into subterranean caverns before spouting out in fountains and geysers.

The existence of clouds depends on the presence of water vapor in the air. So does the comfort of human beings. What matters is not how much moisture the air actually has, but how much additional moisture the air can absorb before it is saturated. We are sensitive to the *relative humidity*, the ratio of the moisture in the air to the moisture it would have if it were saturated. When the relative humidity is 100%, when the air is saturated, the further addition of water vapor leads to condensation. At that point, clouds appear, and complaints

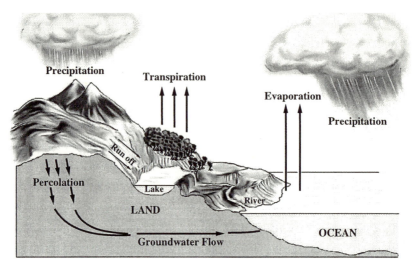

Figure 5.2   The hydrological cycle involves the flow of moisture among reservoirs, principally the oceans (which contain 97% of the water on Earth), glaciers (2.4%), underground reservoirs (.6%), lakes and rivers (.02%), and the atmosphere (.001%).

about the humidity reach a crescendo. If the relative humidity is low, it is possible to be comfortable in spite of high temperatures because the dry air will readily accept moisture that evaporates from our skins. That evaporation keeps humans cool and comfortable. Evaporation when temperatures are low will, of course, make us feel chilly. It may be the reason why, in winter, humans feel cold in a room with low humidity. Humidifying a cold room will inhibit evaporation and keep its occupants warmer.

One way to saturate dry air is to increase its concentration of water vapor. Another way, which does not require the release of more water vapor into the air, is to lower the temperature of the air. Figure 5.3 shows how saturation depends on the temperature and the amount of water vapor in the air. If a volume of air has properties corresponding to point A in figure 5.3, the addition of more water vapor leads in the direction of B, toward saturation and the appearance of clouds. An alternative route to saturation, the move from A to C, is the one we follow when we fill a glass with water and ice on a humid day. The air in contact with the glass is cooled below the dew point temperature, the temperature at which air becomes saturated. The water vapor in the air therefore condenses onto the glass. (The liquid on the

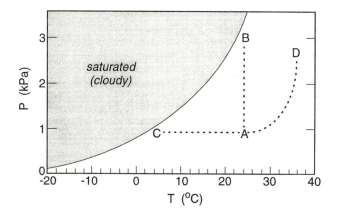

Figure 5.3 The saturation of air with water vapor depends on both its temperature and water vapor pressure (which measures the amount of vapor in the air). To saturate a parcel at A, it can either be moistened (by moving to B) or cooled (by moving to C). For a runaway greenhouse effect, the addition of heat and moisture must result in an unsaturated atmosphere, by moving from A to D, for example. (The vertical axis measures saturation water vapor pressure as explained in Appendix A5.1.)

outside of the glass comes from the air, not from inside the glass). The same process produces both morning dew on the grass—falling temperatures during the night effect a transition from A to C—and frost, when temperatures decrease below the freezing point of water.

If a layer of air above the ground—and not merely the ground itself—is cooled sufficiently during the night, clouds appear as haunting fog. Along the coasts of California and Oregon, sea surface temperatures are lower near the shore than farther offshore. Air moving eastward toward the coast, toward San Francisco, is therefore cooled and becomes saturated, forming fog.

The rate of evaporation from the ocean depends on the relative humidity of the air, which in turn depends on the temperature of the air. This relationship has intriguing implications. If the temperature of the air were to rise steadily so that the air never became saturated—a possible route on figure 5.3 starts at A and proceeds along the dotted line to D—then evaporation would continue until the oceans disappeared. Such a catastrophe is possible because evaporation can create conditions that promote further evaporation. Because water vapor is a greenhouse gas, the more water vapor there is in the atmosphere, the greater is the greenhouse effect and the warmer is the atmosphere. If evaporation causes such a large increase in temperature that the atmosphere remains unsaturated, then evaporation from the ocean into the atmosphere continues, causing the atmosphere to become even warmer, and so on. This is known as a *runaway* greenhouse effect and probably occurred on Venus. On that planet, sunshine is so intense that the greenhouse effect attributable to a certain amount of water vapor in the atmosphere is far larger than it is on Earth. Temperatures are so high on Venus that its atmosphere could not become saturated (see chap. 10.) On Earth, sunlight is less intense, because we are farther from the sun and temperatures are lower, low enough that the atmosphere can become saturated with water vapor, permitting the existence of clouds.

Unsaturated air becomes saturated when it is cooled to its dew point temperature. This occurs when moist air is elevated because temperatures in the atmosphere decrease with increasing height. On a sunny afternoon in summer, the air above the hot ground rises spontaneously when Earth's surface is sufficiently warm, the way water in a pot over a fire rises spontaneously because of convection. A rising parcel of air cools off until, at a height known as the *cloud base*, it is saturated and clouds appear.

Air can be cooled below its dew point temperature without condensation of its water vapor provided the air is absolutely clean and is free of dust and other aerosols. Because the presence of such nuclei is essential for the beginning of condensation and the appearance of clouds, there have been experimental efforts to use airplanes to seed clouds with nuclei to enhance rainfall. Despite numerous attempts, it is still unclear whether such efforts are effective. Banning open fires has been successful in reducing the occurrence of thick fogs over London. (The fog appeared when winds brought water vapor to the soot-laden and therefore nuclei-rich air.) In the atmosphere, especially over the oceans, the nuclei for condensation are often sulfate particles. A major source of such particles appears to be microscopic marine plants. This has led to the speculation that these plants can control Earth's temperature: when too hot, they release more sulfur so that clouds become more abundant and reflect more sunlight, thus cooling Earth.

Convection can cause the appearance of cumulus clouds, which, in their early stages, have sharp boundaries and resemble cauliflowers. The edges soon start to fray and the clouds turn into fluffy cotton-balls. A few of these clouds can, within a matter of hours, grow into threatening, thundering cumulonimbus towers by entraining the surrounding moist air. The latent heat released when the water vapor in the air condenses into water droplets makes the air inside the cloud more buoyant. The rising motion enables the cloud to grow vertically and to entrain more moist air so that the growth of a few cotton puffs into towers is usually accompanied by the demise of the other puffs.

On a warm summer afternoon, clouds readily arise, drift with the wind, disappear, and reappear. They can become menacing and turn into a thunderstorm but are capable of much worse when the elements organize into a swirling, frightening hurricane. Once again, water is a source of energy. The winds that spiral in toward the base of the hurricane reap moisture and latent heat from the ocean. When the swirling air rises in the hurricane, the water vapor condenses and releases the latent heat, stoking the buoyant air to rise more rapidly. This rising motion is sustained when the air over the ocean spirals in faster and faster, harvesting more and more moisture, causing the hurricane to become fiercer and fiercer. This intensification comes to an abrupt halt, and hurricanes start to die once they move over land and are cut off from their source of fuel, the ocean.

The careful orchestration of the release of latent heat that is necessary for a hurricane to develop is possible only over the warm waters of tropical oceans during the season of maximum temperature. Hurricanes become more numerous when the areal extent of those regions increases. This happened in the tropical Pacific during El Niño of 1982 to 1983 when the surface water of the entire tropical Pacific, not only the western side, had high temperatures. On that occasion French Polynesia, noted for its benign climate, was battered by five full-blown hurricanes. Whether global warming will lead to an increase in the number of hurricanes is unclear. An increase in the area covered by warm surface waters will make hurricanes more likely. Greenhouse warming, however, could increase the vertical stability of the atmosphere, inhibit convection, and hence decrease the number of hurricanes.

In a hurricane, the swirling winds that gather moisture (in effect, heat) over a huge oceanic area converge onto a small region where that heat is released. Something similar happens, on a far larger scale, in the case of the monsoons over India. The Tibetan plateau becomes very hot during summer and, when the air over it rises, the winds that converge onto that region pick up moisture from the warm Indian Ocean. This water vapor condenses where the air rises. The release of latent heat makes the air even more buoyant and intensifies the monsoons, ensuring heavy rainfall over India. For similar reasons, rainfall is heavy over the Amazon basin; the trade winds that converge onto it bring moisture from the Atlantic. Much of that moisture is recycled locally: trees efficiently transport water from their roots to their crowns, and by transpiration, to the atmosphere. Destruction of the tropical forests could alter the climate significantly because this considerable recycling will cease, thus decreasing rainfall. Deforestation would also increase the surface albedo—forests are very dark—leading to lower temperatures when less sunlight is absorbed. This could mean weaker convection, which again means less rainfall. The destruction of tropical jungles affects the heat budget of the land by changing both its albedo and its evaporative properties and could result in much drier climates locally. The destruction of any vegetation over a huge area can affect the local climate. The severe dust storms that plagued the central United States from 1934 onward were in part attributable to misguided farming practices. Overgrazing at the edge of the Sahara desert could similarly be contributing to its expansion.

When air rises spontaneously, when the atmosphere is unstable, cumulus clouds appear as playful cotton puffs, threatening cumulus towers, or, in the tropics, a fearsome hurricane. When air is forced to rise because of a mountain, clouds loiter around the peaks and provide rainfall or snow, primarily on the windward side. Waves in the lee of a mountain permit clouds to settle along their crests where the air rises. Although the clouds themselves remain stationary, there is a continual flow of air through them; new water vapor condenses at one end and evaporates at the other end. Similarly, when clouds hover over an isolated warm island where the air is rising, there is continual condensation and evaporation. The situation is analogous to a frothy, foaming waterfall that is stationary even though there is a continual flow of water through it.

Warm and Cold Fronts

Clouds are locked in place only near mountains and over warm, isolated islands. Fortunately, moist air is elevated in many other ways, including maneuvers of huge air masses over hundreds of thousands of square kilometers. For example, when warm moist tropical air from the Gulf of Mexico charges northward, it soon encounters cold dense air that digs in, forcing the warm air to rise and spread out in layers. In the vanguard of such an approaching warm front are the very high, fleecy cirrus clouds:

> Like the bright hair uplifted from the head
> Of some fierce Maenad, even from the dim verge
> Of the horizon to the Zenith's height,
> The locks of the approaching storm.
>
> (Percy Bysshe Shelley, "Ode to the West Wind")

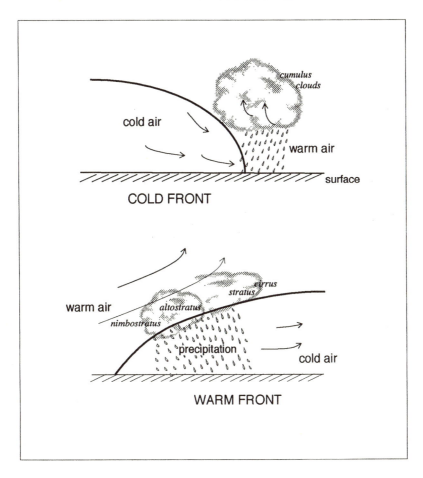

Figure 5.4 Idealized warm and cold fronts that involve, respectively, the movements of air masses that are moist and warm, and cold and dry.

Those locks lend the moon a halo and signal the approach of stormy weather. A warm front can be anticipated because it leans forward (see fig. 5.4). First come the very high, fleecy cirrus, then a lower sky of solid webs, the cirrostratus, followed by the even lower altostratus, which indicate that rain is not far off. Finally the arrival of nimbostratus signals the start of a long slow rainfall. A cold front marks the clash between cold, dense air, which has gone on the offensive, and warmer, less dense air. In this stormy affair, warm air is forced upward so that cumulus clouds enter the fray.

Fronts, warm or cold, usually mean rain. Stratus clouds, especially when thin and flat, are capable only of drizzles of small rain drops. In tall cumulus towers, however, water droplets travel up and down and through accretion can grow substantially in size before exiting. That is why cumulus clouds can produce heavy rains, even cloudbursts. When a cumulus cloud is tall enough for temperatures in its crown to be well below freezing, then droplets that reach those altitudes turn into ice. Repeated visits to the crown create hailstones that can grow to the size of golf balls before they are so heavy that they overcome the updrafts and fall to the ground. (A record-sized hailstone with a weight of nearly 11 pounds fell in the Guangxi region of China on May 1, 1986.) In winter, in mid and high latitudes, cumulus clouds do not have to be particularly tall for temperatures at the crown to be below freezing. Under such conditions, water droplets do not automatically turn into ice but can become supercooled. Minute ice nuclei must be present for ice crystals to form, and those can grow by accretion when crystals collide with supercooled droplets. Ice crystals are delicate and fracture or splinter easily during collisions, thus creating more ice particles onto which supercooled water can freeze. The result is a cascade, producing many ice crystals that sometimes stick together to form snowflakes in a variety of shapes (see fig. 5.5).

Snow readily forms inside clouds. Whether any snow reaches the ground depends on the temperatures below the cloud. If those temperatures are too high—that is usually the case in summer—then the snow melts on its way down. If temperatures are so low that the melting is only partial, then the thin film of water that surrounds a flake acts like glue that can join several flakes to form gigantic ones. Temperatures near freezing and moist air above the ground are the conditions that favor these soggy flakes. For small, powdery flakes, the air near the ground must be very cold and low in moisture.

Sleet and freezing rain are possible when temperatures vary as in figure 5.6. As a snowflake travels downward from the base of a cloud, it encounters a layer of warm air where it melts and turns into a supercooled droplet. This droplet freezes in the very cold air above Earth's surface and becomes an ice pellet that makes an audible sound when it strikes a window or other hard surface. This form of precipitation is known as sleet. If the layer of cold air near the surface is so shallow that the liquid droplet remains supercooled as it travels through the layer, then, upon striking a cold surface, the droplet spreads out, freezes, and glazes that surface. The result is a wondrous landscape in which everything is coated with a shiny layer of ice. It is both beautiful and dangerous; roads become extremely slippery, and the weight of ice can break tree branches, telephone cables, and power lines.

Figure 5.5 A few examples of the shapes snowflakes can assume. From Bentley and Humphreys (1931).

Clouds are the principal source of uncertainty in estimates of future global warming caused by higher atmospheric carbon dioxide levels. If more carbon dioxide increases temperatures, and hence increases evaporation from the ocean, then water vapor can magnify the warming because it is a powerful greenhouse gas. On the other hand, if there is condensation and clouds become more abundant, then they will reflect more sunlight and cause cooling. Whether or not the warming exceeds the cooling depends on the cloud form—cumulus, stratus or cirrus—and on factors such as the size of the droplets. Clouds composed of a huge number of small droplets scatter sunlight more effectively than clouds with a small number of large droplets. A photon from the Sun changes direction when it strikes a droplet. The more often a photon changes direction after it starts traveling downward from the top of a cloud, the more likely it is that the photon will reemerge from the top. Hence, thick clouds with a great many tiny droplets reflect a considerable amount of sunlight and appear very bright when seen from space and very dark when seen from the ground. Such clouds cool the planet. A cloud can also cause warming, depending on temperatures at its base, where earthlight is absorbed,

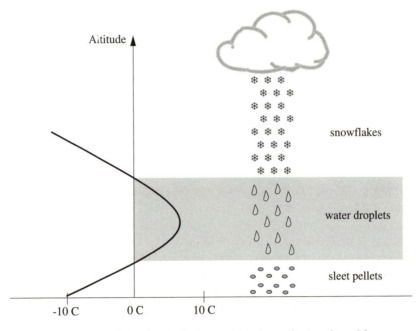

Figure 5.6   A snowflake that falls from a cloud can first melt and become a water droplet when it encounters a warm layer of air and then freeze and become sleet when it falls through a cold layer of air.

and at its crown, where heat radiates to space. Tall cumulus towers with a low, warm base and a cold, tall top keep Earth's surface warm but cover a relatively small area. Stratus clouds can cover a huge area but are often only a thin sheet. Computer models that attempt to estimate future global warming must simulate these various shapes that change continually. Imagine computer models as whimsical as clouds!

6

THE CLIMATE TAPESTRY

THE COLORFUL TAPESTRY created by Earth's rich diversity of climatic zones depends primarily on a colorless dye, water. Its presence turns the landscape green (or white, when it falls as snow); in its absence, yellows, browns, and rich oriental colors prevail. The winds harvest their dye over the oceans and deposit their crop in fantastically shaped granaries, clouds that range from tall cumulus towers to vast sheets of stratus. When and where it rains depends on the winds, which hold the secret to the design of the tapestry.

The winds do more than determine Earth's climate. In the thirteenth century, Bartholomew, a Franciscan friar, explained that "men in the north are tall in stature and fair in body" because the north wind dries out and cools the land, and closes the pores of the body which then retains heat better. The south wind, which is hot and moist, has the opposite effect so that "men of the southern lands are different from those of the north in stature and appearance. They are not so bold nor so wrathful." A combination of these winds could presumably produce the ideal climate and people. Early in the twentieth century, the Yale geographer Ellsworth Huntington tried to determine the climatic conditions most conducive to physical and mental well-being. From analyses that included the efficiency of 65 young women in a label-pasting factory in North Carolina and the mathematics grades of 240 cadets at West Point, he concluded that an average temperature of 64°F (18°C), a relative humidity of 60%, and considerable daily and seasonal weather fluctuations, are close to ideal. He also wrote that no nation "has risen to the highest grade of civilization except in regions where the climatic stimulus is great," and proposed that changing climates contributed to the decline of the civilizations of Egypt, Persia, and Rome.

Today, although few people agree with Huntington's conclusions, few deny that climate influences culture. Culture, in turn, influences our perception of natural phenomena. An interesting example is the change, over the past century, in our perception of the phenomenon known as El Niño (for a detailed discussion, see chap. 9). This temporary change in the climate of the tropics, especially that of the Pacific, was at first celebrated as a joyous event along the normally arid coasts of Ecuador and Peru; El Niño brings rains to that region and transforms the desert into a garden. That still happens today but we are now more conscious of the severe storms and torrential rains that accompany El Niño, to such an extent that we now view El Niño as an unmitigated disaster. The character of El Niño has not changed, but the people of Ecuador and Peru have become very vulnerable to fluctuations in climatic conditions. Rapid growth in population has been accompanied by a huge increase in the number of houses, roads and bridges that can be destroyed by severe floods. A change in social conditions has caused a change in our perception of a climate fluctuation.

Climate influences culture, and culture in turn influences our perception of climate. The close ties between climate and culture create the impression that, because the one is a regional phenomenon, the other is, too. In reality, this planet's different climatic zones are all related by the winds. These invisible threads of the climate tapestry weave the deserts and jungles, the steppes and the tundra, into a cohesive whole. The winds also cause the tapestry to be vulnerable. Because of them, a tug in one place is felt throughout; the entire tapestry is threatened if any part of the fabric starts to unravel. That is why the preservation of a tropical forest is of benefit to all; why pollution anywhere is a problem everywhere. Because of the winds, there are no true islands on this planet.

Mountain Winds

The mistral that roars down the Rhone valley, the bora of the northern Adriatic, the chinook and foehn of the Rockies and Alps, respectively, are all mountain winds driven by a force that is simulated in the U-tube shown in figure 6.1a. In this tube, the pressure at point P is clearly greater than the pressure at O. A pressure force from P to O therefore drives the water in the direction of the arrow. A similar force can drive air through a gap in a mountain range that separates dense cold air from warm moist air. If there is cold, dense air over a high plateau, and warm moist air over low land or over the adjacent ocean, then the wind rushes through any gaps (passes) in the mountain, accelerating as it falls. Mountain winds such as the chinooks attain high speeds in this manner. Descent compresses the air, because pressure increases with decreasing altitude, causing a rise in temperature. That temperature increase explains why mountain winds such as the Santa Ana in southern California are hot and dry.

(a)

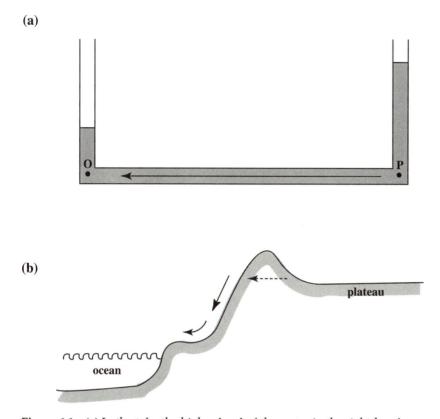

(b)

Figure 6.1 (a) In the tube the higher level of the water in the right-hand arm translates into a pressure that is higher at P than at O so that the water flows in the direction of the arrow. (b) Such a pressure force drives the winds from a high, cold plateau, where pressure is high, through a mountain pass or down a valley toward a region of lower pressure.

Sea Breezes

The refreshing sea breezes that blow from the ocean toward the adjacent warm land on a sunny day in summer are so dependable in some regions that they have earned names: the virazon of Chile, the imbat of Morocco, the ponente of Italy, and the kapalilua of Hawaii. These breezes are most intense in the afternoon when the land attains a temperature much higher than that of the adjacent ocean. The air in contact with the land becomes so hot that it rises, to be replaced by cooler air that moves in from the ocean. Aloft, motion is in the oppo-

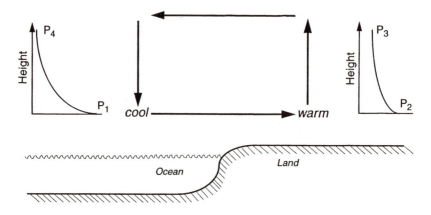

Figure 6.2 Over the hot land, air molecules are energetic and rise to far greater heights than they do over the cold ocean. Pressure aloft is therefore higher over the land than over the ocean—P_3 is greater than P_4—so that the air aloft flows from the land toward the ocean. At the surface, the situation is reversed—P_2 is smaller than P_1—and the flow is a sea breeze, from the ocean toward the land.

site direction, from the land to the ocean. Offshore, the air descends and closes the circuit.

The temperature difference between the ocean and land creates a pressure force that drives the sea breeze. The molecules of the hot air over the land are so energetic that they defy gravity far more easily than do the molecules of the cool air over the ocean. Thus, the density of the air, and hence the pressure, decreases far more slowly with elevation over the land than over the ocean, as seen in figure 6.2. As a result, the pressure force aloft, from the region of high pressure to the region of low pressure, is from the land toward the ocean. Near the surface, the flow of air is in the opposite direction.

Monsoons

Monsoons are similar to sea breezes but are seasonal rather than diurnal. Southwesterly winds bring moisture from the ocean toward the Indian subcontinent in summer when the land is hot and the ocean less so. In winter, the winds reverse direction and blow from the cold continent toward the relatively warm ocean. The moisture that the winds carry is of enormous importance. Air that rises over the land is buoyant, not only because of the heat it acquires from the hot land, but also because of the latent heat it acquires when the water vapor in the air condenses to form clouds. This additional buoyancy intensifies the rising motion and hence the winds that converge onto the land. Those winds bring yet more moisture, acquired from the ocean. The monsoons, in effect, operate like a heat engine by taking heat from the ocean, where liquid water is converted into water vapor during evaporation, and by depositing that heat in the atmosphere above the land where the water vapor is converted back into the water droplets of clouds. The ocean recovers the heat from the sun.

To explain monsoons as simply magnified sea breezes, is to be ignorant of the following facts: India can be much hotter in May before the onset of the monsoons than in July when the monsoons are most intense; the hottest part of India, the northwest, gets no rain during the monsoons; the average temperature of India can be greater during years of poor rainfall than years of good rains. The monsoons clearly depend on more than the temperature difference between the Indian subcontinent and the adjacent seas. It is part of the global atmospheric circulation and, as such, can be affected by changes in remote regions. Of paramount importance to the global circulation, and the links it creates between different regions, is Earth's rotation. We next explore why the rotation causes the salient features of the global circulation to be, not gigantic sea breezes from the warm equator to the cold poles, but Jet Streams, one in each hemisphere, that blow from west to east.

The Hadley Circulation

Christopher Columbus proposed to reach the east by sailing to the west. When he left Portugal, however, he set sail for the southwest, not the west, because he wanted to take advantage of the easterly trade winds that prevail in the tropics. On his return journey from the New World, he took a northerly route to take advantage of the westerly winds that prevail in this planet's middle latitudes. Between the zones of westerly and easterly winds are the calm, dry, sunny horse latitudes, where Coleridge's Ancient Mariner presumably got stuck

> Day after day, day after day,
> We stuck, nor breath nor motion;
> As idle as a painted ship
> Upon a painted ocean.
> Water, water, every where,
> And all the boards did shrink;
> Water, water, every where,
> Nor any drop to drink
> (*Samuel Taylor Coleridge, "The Rime of the Ancient Mariner"*)

The term "horse latitudes" probably came about when sailors encountering these conditions while voyaging from old to new Spain were sometimes forced to jettison the horses they were transporting. That hot, dry climatic zone is in striking contrast to the overcast, rainy doldrums near the equator.

Why are there westerlies in some latitudes, easterlies in others? Those who live in midlatitudes take their prevailing winds, the westerlies, for granted and are surprised to learn about easterlies in the tropics. The wind reports filed by early explorers stimulated numerous speculations about the curious trade winds. (The name is unrelated to commerce; it reflects the steadiness of the winds and has its origin in the word tread, the steady path that a ship follows.) To explain winds that are westward in low latitudes, some natural philosophers proposed that because of its lightness, the warm tropical air is unable to keep up with Earth's surface in its rotation from west to east. Others pictured the winds as exhalation from the sargassum weed in the subtropical oceans. In 1735 a London lawyer, George Hadley, proposed a solution that by and large is correct.

The arguments that explain a sea breeze imply that the latitudinal distribution of solar heating should cause warm, light air to rise near the equator, cold air to sink near the poles. Equatorward motion near Earth's surface and poleward flow aloft should close the circuit, as in figure 6.3. A circulation that resembles this is observed on Venus but not on Earth. The salient features of the atmospheric circulation on our planet are not north-south winds but the intense west-east jet streams that circle the globe, one in the midlatitudes of each hemisphere. Those jets are so rapid that they can accelerate eastward flights, from California to New York and from New York to Europe, for example, while retarding flights in the opposite direction. The principal reason for jet streams on Earth but not on Venus is the different rates of rotation of the two planets: Earth spins about its axis once a day, Venus about its axis once every 243 Earth days. (On Venus a day is longer than a year.) Hadley realized that the rotation of the Earth will affect north-south motion profoundly.

Earth spins from west to east about an axis through its poles. Air at the equator is farthest from the axis of rotation and, because it moves with the solid Earth beneath it, has a much higher eastward speed than air at higher latitudes. It follows that a parcel of air that starts at the equator and travels poleward without changing its eastward speed will move progressively farther eastward relative to the ground beneath it. Viewed from the ground, the poleward-moving parcel is seen to travel eastward, faster and faster as its latitude increases. By the time it reaches Earth's middle latitudes, the eastward flow is so

North Pole

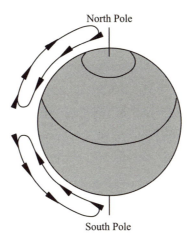

South Pole

Figure 6.3 On a nonrotating planet, air that rises in low latitudes flows poleward aloft, sinks in high latitudes, and returns equatorward near the surface.

intense that it is a swift river of air, a jet stream. Its presence is felt even at Earth's surface; in the latitudes where that current of air prevails, the winds are so tempestuous that they are known (in the southern hemisphere) as the roaring forties and screaming fifties. (Although Hadley proposed that parcels of air conserve their east-west speed as they move poleward, they actually conserve their angular momentum; see Appendix 6.)

Parcels of air that move poleward start with a high eastward speed and hence become intense westerly winds—winds from the west—in midlatitudes. Parcels of air that move from the poles toward the equator—assume that they start with no eastward speed—find that the ground beneath them is moving faster and faster in an eastward direction as they approach the equator. Seen from the ground the parcels appear to be moving westward; they become easterly trade winds.

We on Earth's surface are usually not conscious of Earth's rotation. We can pretend that there is no rotation and can explain the deflection of moving particles caused by the rotation—eastward if they move poleward, westward if they move toward the equator—by invoking a force, known as the Coriolis force. In the northern hemisphere, it deflects particles to the right of the direction in which they are moving; in the southern hemisphere, the deflection is to the left (fig. 6.4). At the equator, the Coriolis force vanishes. Note that this force applies not only to north-south but also to east-west motion as explained in Appendix 6.

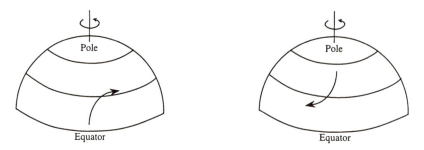

Figure 6.4 *Left,* Parcels of air in low latitudes are initially far from Earth's axis of rotation and therefore are moving rapidly eastward. When they move poleward and retain their angular momentum, they move eastward relative to Earth's surface beneath them. *Right,* Equatorward-moving parcels, on the other hand, move westward relative to the surface beneath them.

The Coriolis force (or the rotation of Earth) has a most remarkable effect on atmospheric motion: it causes that motion to be, not from high toward low pressure as in the case of a sea breeze, but along lines of constant pressure. Initially, motion is in the direction of the pressure force as shown in figure 6.5, but the persistent Coriolis force deflects air parcels to the right until the Coriolis and pressure forces balance each other. At that stage, the wind is blowing parallel to lines of constant pressure (isobars). If the journey of an air parcel traveling at a relatively low speed is of short duration, a matter of minutes, then the effect of the Coriolis force is slight and motion is essentially from the region of high pressure to the region of low pressure. Such is the case with a sea breeze. However, if the air parcels travel for a sufficiently prolonged period, then the Coriolis and pressure forces are of equal importance, which is why large-scale atmospheric motion approximately follows isobars. This result is of enormous importance to weather forecasters, because it permits them to infer winds from maps of the surface pressure. (Such maps require pressure measurements over a large area and a means of communicating that information to a central location. That is why the invention of the telegraph in the 1850s was of great importance to meteorology.)

Let us return to parcels of air that rise near the equator and then travel poleward aloft. If they conserve their angular momentum as they move, then their eastward speed increases so rapidly that, by the time they reach midlatitudes, they are traveling at the speed of sound! In reality, other processes come into play long before this happens. (The next chapter describes how the Jet Streams start to meander

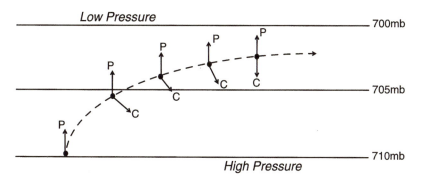

Figure 6.5 The rotation of Earth dramatically alters atmospheric motion in response to pressure differences. Initially a pressure force, P, accelerates air parcels from areas of high pressure toward areas of low pressure, but the Coriolis force, C, persistently deflects air parcels to the right (in the northern hemisphere). Ultimately, the parcels move in such a direction that the pressure and Coriolis forces are equal and opposite; at that stage, the wind blows, not from high pressure toward low pressure, but along lines of constant pressure.

spontaneously, causing various time-dependent phenomena associated with weather.) Before the air aloft reaches unrealistically high speeds, much of it sinks to the surface in the neighborhood of the latitude bands 30°N and S. Some of that air returns equatorward near the surface, is deflected to the west by the Coriolis force and hence becomes the easterly trade winds. This tropical cell, in which air rises at the equator, flows poleward aloft, subsides near 30° latitude, and returns to low latitudes along the surface, is known as the Hadley circulation.

Some of the air that sinks in the neighborhood of 30° latitude spreads poleward, as shown in figure 6.6, and is deflected eastward by the Coriolis force. This flow, plus the surface expression of the intense Jet Streams aloft, are responsible for the westerly winds in midlatitudes. In the neighborhood of 60° N and S, the poleward-moving air at Earth's surface encounters air moving toward the equator that subsided over the cold poles. The atmospheric circulation between the equator and pole is therefore broken into a number of cells as shown in figure 6.6.

The temperature difference between the equator and the poles, and the rotation of Earth determine the circulation in figure 6.6 and hence determine the locations of Earth's principal climatic zones. The air that moves to lower latitudes near the surface gains moisture from the ocean and releases it as rainfall in the cloudy doldrums where the

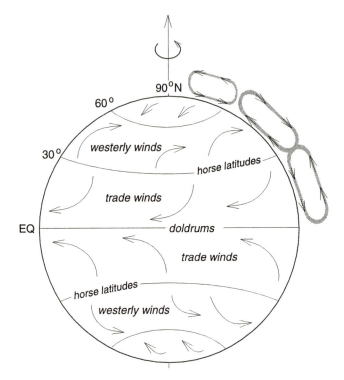

Figure 6.6   The southeast and northeast trade winds in the tropics converge onto the doldrums, rainy regions where moist air rises. In the subtropics, the easterly and westerly winds diverge from the horse latitudes, sunny, dry regions over which dry air subsides. The easterly winds in polar regions, and the westerlies in subpolar regions, converge onto rainy regions near 60°N and 60°S, respectively.

air rises, which is why hot, steamy jungles are located in low latitudes. In the upper atmosphere, the air, which is now drained of its moisture, diverges from the equator and proceeds to the subtropics, where it sinks over the horse latitudes. Descent causes compression and heating (see chap. 4) so that the air has a low relative humidity. For this reason, Earth's principal deserts—the Australian, Sahara, and California deserts, for example—are located in the subtropics. These deserts and the equatorial jungles are related in the same way that the arid lee side of a mountain is related to its verdurous windward side: the air that subsides over the dry regions lost its moisture earlier, when it rose over the rainy region.

Cities such as Los Angeles and Phoenix have severe local air pollution problems, in part because of their location in the horse latitudes where subsiding dry air prevents polluted air from rising. (Additional factors that play a role include the local terrain.) All things being equal, Seattle is unlikely to have pollution as severe as that of Los Angeles because at Earth's surface, the air converges onto the latitudes of Seattle, causing the air to rise, dispersing pollutants and bringing plentiful rainfall.

The Effects of Land-Sea Contrasts

The atmospheric circulation described thus far is unrealistic in creating climatic zones that vary only with latitude. A planet without continents would have such a climate. On Earth, climate can vary significantly along a circle of latitude. Near the equator, for example, New Guinea has lush jungles, whereas the Galapagos Islands and the coastal zone of Ecuador are practically barren. San Francisco, Denver, and Washington, D.C., all along 35°N approximately, have very different climates. The reasons for such differences are varied: the different thermal properties of continents and oceans, mountains that run north-south, and complex sea surface temperature patterns.

The salient features of the atmospheric circulation in midlatitudes are the jet streams that circle Earth. The effect of continents on those currents of air are at a minimum in the southern hemisphere, because that region has practically no continents; the coast line of Antarctica happens to be close to a circle of longitude so that it hardly interferes with east-west winds. Therefore, on a map of pressure at Earth's surface (fig. 6.7), conditions are closest to those of a water-covered planet in high southern latitudes. Matters are very different in high northern latitudes where continents with tall mountains cover much of Earth's surface. They influence the path of the jet stream, especially in winter when the ocean is warm relative to the very cold land. The results include prominent low-pressure zones, centered on Iceland and the Aleutian Islands, on a map of surface pressure in winter. Those zones are usually overcast and rainy because there is a tendency for the surface flow of air to converge onto them, causing rising motion and condensation of water vapor. To understand why this is so, keep in mind the tendency of parcels of air to move along lines of constant pressure so that they circle a center of low pressure as shown in figure 6.8. The term *cyclone* refers to anticlockwise motion around center of low pressure. (A *hurricane* or *typhoon* is a tropical cyclone.) In an anticyclone motion is clockwise (in the northern hemisphere) around a center of high pressure (fig. 6.8).

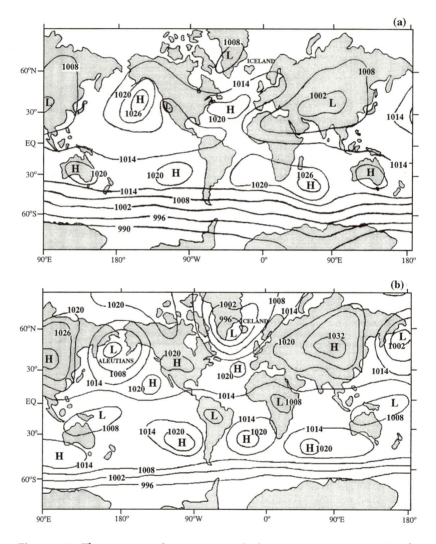

Figure 6.7 The presence of continents and of mountain ranges modifies the circulation of a water-covered globe shown in figure 6.6 by introducing longitudinal variations. In surface pressure (measured in millibars [mb] on the maps), they take the form of prominent centers of low pressure and high pressure over different parts of the globe. (a) Conditions in July. (b) Conditions in January when Iceland and the Aleutian Islands have foul weather associated with centers of low pressure that persist during winter. In summer, the eastern parts of the subtropical oceans have fair weather associated with centers of high pressure.

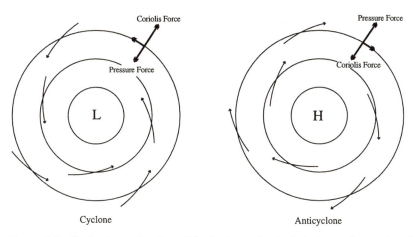

Figure 6.8 Cyclonic motion is anticlockwise and spirals in toward a center of low pressure in the northern hemisphere. Anticyclonic motion is clockwise and spirals outward from a center of high pressure in the northern hemisphere.

In the absence of friction, air parcels moving along an isobar experience two balancing forces: a pressure force and a Coriolis force. Drag is important near the surface of Earth so that a third force, associated with friction that retards air parcels, comes into play. This force leaves the pressure force unaffected but influences the Coriolis force, because it slows parcels down and the Coriolis force is proportional to the speed of the parcels. It follows that, near Earth's surface, the Coriolis force is unable to balance the pressure force. As a consequence, the air parcels near the surface spiral inward toward the center of low pressure, where the air is forced to rise. Rising air expands and cools so that the moisture in the air condenses into clouds and rain is then likely. Therefore, low pressure is often associated with overcast, rainy conditions, and a barometer indicates rainy weather when pressure falls.

Whereas conditions in the extratropics depend critically on the intensity and paths of the Jet Streams, conditions in the tropics are determined primarily by temperature patterns at Earth's surface. On a water-covered globe, air rises near the equator where surface temperatures have a maximum and sink in the horse latitudes as in Hadley's idealized model. On Earth, this is almost the case over the Pacific and Atlantic Oceans, where the effects of continents are minimal. Over those oceans, rising motion is confined to the regions of very high sea-surface temperature that are seen to be just to the north of the equator (fig. 6.9). The southeast and northeast trade winds con-

Figure 6.9 Sea surface temperature (in °C) varies not only with latitude but also with longitude. Along the equator in the Atlantic and Pacific Oceans, temperature increases significantly from east to west. These sea surface temperature patterns have a profound effect on the atmospheric circulation because, in the tropics, convection occurs over the regions with the highest surface temperatures. Temperatures exceed 28°C in shaded areas.

verge onto these doldrums, which are cloudy zones of heavy rainfall. In addition to these east-west bands of clouds over the oceans, massive banks of cumulus clouds tower over the other major regions of high surface temperatures: the basins of the Amazon and Congo Rivers, and the "maritime" continent of the warm western tropical Pacific, southeastern Asia, and northern Australia. The surface winds converge onto these centers of low pressure where the air rises. The Tibetan plateau in summer is also a center of low pressure, as is evident in figure 6.7, in part because of its considerable elevation. (The air over the plateau has a low density and, when heated, readily rises.) Scientists believe that, during the absence of that plateau—it appeared only after India collided with the Asian continent millions of years ago—the monsoonal winds that converge onto the Tibetan low pressure center were far weaker than they are today.

The air that rises over the centers of low pressure in the tropics where surface temperatures are at a maximum, sinks over the adjacent regions with relatively low temperatures, high surface pressure, and low rainfall. The prominent subtropical high pressure zones in figure 6.7—those off the western coasts of northwestern and southwestern Africa, California, Chile, and Australia—therefore have links to the low-pressure zones over the adjacent continents. It has been suggested, for example, that the air that rises over the Tibetan Plateau in summer tends to sink over the eastern Mediterranean and part of the Sahara, thus contributing to the low rainfall of those regions.

Temperature differences at Earth's surface exist not only between land and sea but also within the oceans, between the eastern and western tropical Pacific, for example. Figure 6.9 shows that the waters bathing the coasts of Ecuador or Peru can be as much as 10°C colder than the waters, at the same latitudes, along the shores of Indonesia and northern Australia. This huge temperature difference across the tropical Pacific Ocean induces an atmospheric circulation similar to the monsoons: air rises over the warm waters of the western Pacific and subsides over the cold waters of the eastern Pacific. This Walker circulation, as it is known, is closed by easterly trade winds near the surface, westerly winds aloft. The subsiding air is so dry that a large part of the eastern tropical Pacific receives little rainfall. (The subtropical deserts such as those of California, the coast of Peru, and the Sahara in effect extend far westward over the adjacent oceans.)

Whereas cumulus towers form over the tropical regions with high surface temperatures, very different low, stratus clouds that produce essentially no rain, cover the cold waters off the coasts of Peru, California, and western Africa. Because the air subsides over those regions, the moisture that evaporates locally from the ocean is unable to rise into towers and instead forms stratus clouds. Those who visit Lima, the capitol of Peru, for the first time are inclined to believe that, because the clouds are low and look threatening, rain is imminent. The clouds are perennial, but it very seldom rains in Lima.

In the tropics, the temperature pattern at Earth's surface practically determines the atmospheric circulation. The seasonal change in that pattern causes the seasonal movements of the tropical convective zones of heavy rainfall. Over the continents, those zones essentially "follow" the Sun into the summer hemisphere, but, as is evident in figure 6.10, matters are more complicated over the ocean. (Appendix 9 discusses the climatic symmetry relative to the equator in the eastern Pacific and Atlantic, where the convection is mostly north of the equator throughout the year.) Interannual variations in sea surface temperature patterns cause changes in the atmospheric circulation and hence in rainfall patterns that can bring droughts to regions that otherwise have plentiful rainfall, and vice versa. A dramatic example of such a change is El Niño (discussed in chap. 9).

A time-independent atmospheric circulation, one without weather, is possible in principle; it is conceivable that there are planets where atmospheric conditions are the same from one day to the next. That is not the case on our particular planet. On Earth, the chaotic fluctuations in atmospheric conditions known as weather form an essential aspect of the atmospheric circulation and contribute significantly to the global redistribution of heat and moisture, especially in the latitudes of the Jet Streams where westerly winds prevail. Weather (the topic of the next chapter) plays a central role in the design of the climate tapestry.

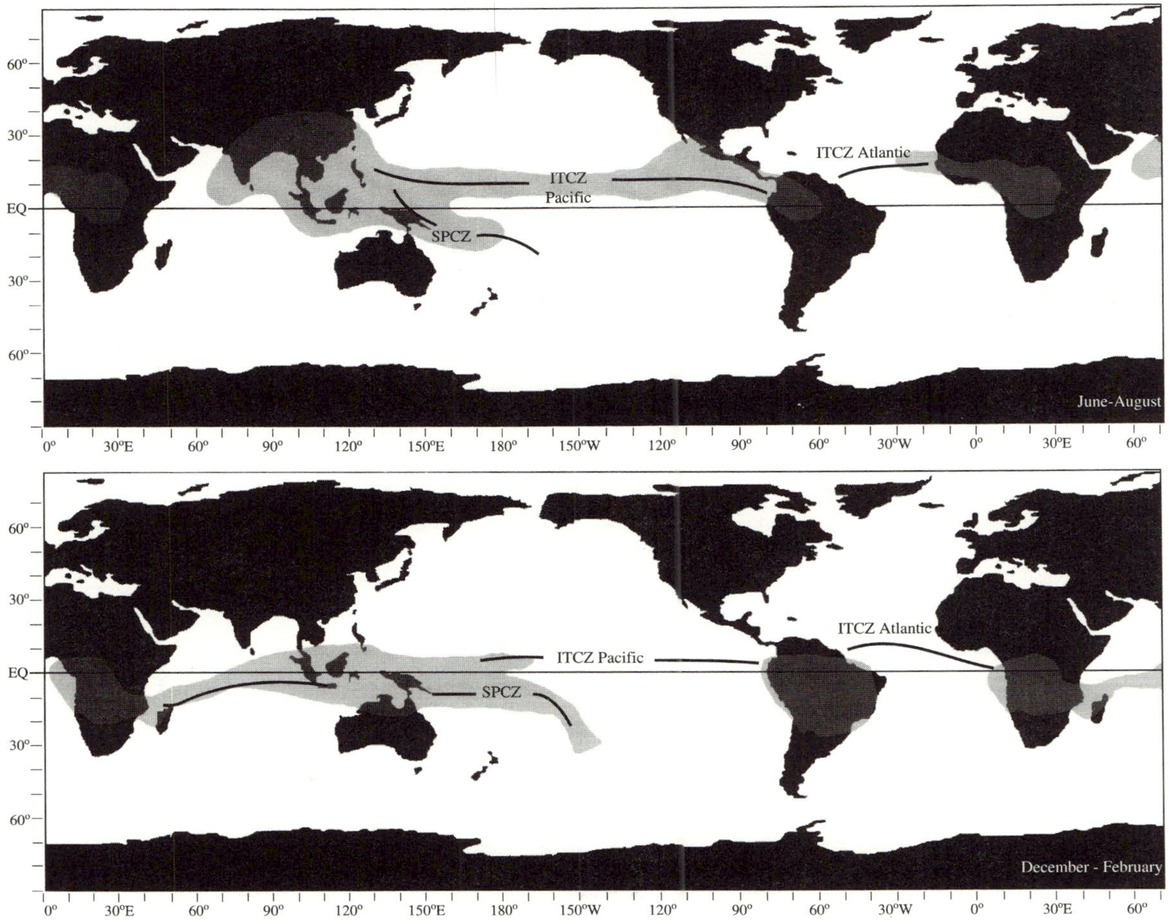

Figure 6.10 Seasonal movements of tropical convective cloudbands that produce heavy rainfall. The doldrums, onto which the southeast and northeast trade winds converge, are denoted by ITCZ for intertropical convergence zone. SPCZ denotes the South Pacific Convergence Zone. The top panel shows average conditions for June, July, and August; the bottom panel for December, January, and February.

7

WEATHER, THE MUSIC OF OUR SPHERE

AFTER A FESTIVE literary dinner in his studio, the painter Benjamin Haydon wrote in his diary for December 28, 1817 that "Wordsworth was in fine cue, and we had a glorious set-to—on Homer, Shakespeare, Milton and Virgil. Lamb got exceedingly merry and exquisitely witty. He then, in a strain of humor beyond description, abused me for putting Newton's head into my picture; 'a fellow,' said he, 'who believed nothing unless it was as clear as the three sides of a triangle.' And then he and Keats agreed that [Newton] had destroyed all the poetry of the rainbow by reducing it to its prismatic colours."

The romantic poets had grown disillusioned with the "hard" science of Newton, the bearer of light for the Age of Reason. They perceived the Enlightenment, with its emphasis on the universal and timeless, as mechanistic and lacking in emotion. Their reaction, in favor of genius, inspiration and diversity, led to a preference for the "soft" life sciences over the hard physical sciences. (Wordsworth, Shelley, and Coleridge all owned microscopes; Keats spent six years as a medical apprentice and qualified as an apothecary.)

Today the distinction between the hard and soft sciences is blurred. The study of phenomena such as weather amounts to a marriage of the soft and hard sciences; it involves exact equations that describe, not a predictable world, but a chaotic one that has infinite variety. Scientists know the laws that govern atmospheric winds and are able to make valuable inferences from those laws, but they nonetheless are very limited in their ability to predict the behavior of those winds. Consider, for example, the wild west wind of Shelley's poem:

> O wild west wind, thou breath of autumn's being
> Thou, from whose unseen presence the leaves dead
> Are driven, like ghosts from an enchanter fleeing
> *(Percy Bysshe Shelley, "Ode to the West Wind")*

Why are the winds wild? Why especially so in autumn and winter? Newton's laws, as scientists discovered in the twentieth century, provide answers to these questions, thus enhancing our appreciation of

Shelley's lines. Those laws can furthermore be used to make reasonably accurate forecasts of the weather for a limited period. However, they do not enable scientists to predict precisely in what way the winds will be wild; occasionally weather forecasts are wrong. This mixed success evokes an ambivalent response: many people dismiss weather forecasts as hopelessly inaccurate but nonetheless follow them closely, to such an extent that weather is a daily feature of both television and newspapers and even has a television channel entirely devoted to it, 24 hours a day. Weather provides everyone with an opportunity to occupy two positions at the same time, like a photon that is both a particle and a wave. By paying close attention to the results from mathematical models of the atmosphere, it is possible to obtain useful information about evolving weather patterns and also reassuring confirmation of the romantic belief that nature is capricious and unpredictable. That belief attributes inaccurate forecasts to experts at a centralized bureaucracy (the Weather Service), who prefer the perspective from a polar orbiting satellite to "looking out the window." To predict "our" weather, those experts rely too much on computer models of the global atmosphere.

"Our" weather can indeed be very different from that in other regions. An English lady made this discovery while on vacation in southern California. Upon opening her window one morning she was heard to complain about "another bloody sunny day." The lady was growing tired of predictable warm days followed by predictable cool evenings. To her, the daily fluctuations in atmospheric conditions in southern California are as monotonous as the music of a pendulum or a tuning fork. The lady longed for additional notes; weather that changes continually so that each day is distinct. The music in England, rather than that of a tuning fork, is the music of an instrument capable of a tune. English weather is infinitely varied and has recurrent themes or leitmotivs that can be learned by watching the sky carefully: a ring around the moon is frequently the precursor of stormy weather, red skies at night are usually a shepherd's delight— the next day is likely to be splendid.

The fluctuations in atmospheric conditions over southern California and England represent, respectively, the forced and natural variability of the atmosphere. The forced variability is related to the daily and seasonal changes in sunlight in a direct and obvious manner. Natural variability is more subtle and would be present even if there were no changes in sunlight. Its relation to music is more than metaphorical; the weather in places such as England can literally be viewed as the music of the atmosphere. In the same way that a musical instrument has modes of oscillation that are easily excited—by plucking a violin string or blowing into a flute, for example—so the spherical shell of gases that envelopes our planet has natural modes of oscillation. The counterpart of a taut violin string is a swift atmospheric wind or jet that vibrates and undulates to produce the counterparts of musical notes: cyclones, fronts, and various other phenomena associated with weather. Where the atmospheric winds, aloft rather than at the surface, are particularly intense—the latitudes of England, for example— natural variability is prominent and weather is fascinating.

Weather refers to the daily fluctuations in atmospheric conditions. Climate, the long-term average of the daily variations, can also fluctuate. If the English lady had stayed in southern California sufficiently long, she would have discovered that, in that region, the most prominent variability is not the day-to-day change in weather, but longer-term climatic fluctuations. Some winters are drier and cooler than others; prolonged droughts alternate with periods of plentiful rainfall, even disastrous floods. If the weather of England were the music of a high-pitched violin or flute, then the climate fluctuations of California bring to mind a cello or bassoon. Other parts of the globe—India, where the monsoons fail occasionally, and the tropical Pacific which has a Southern Oscillation between warm, wet El Niño and cold, dry La Niña—call for more instruments. Not a tuning fork, not an instrument capable of a tune, only a huge symphony orchestra can do justice to the music of this planet.

The atmosphere on its own can produce the music we call weather, but to produce climatic fluctuations, it needs to cooperate with the land, water, and ice surfaces beneath it. This cooperation is possible because atmospheric winds both depend on and influence temperature patterns at Earth's surface. For example, over the oceans, the winds drive currents, such as the Gulf Stream, which transport huge amounts of warm water poleward, thus creating surface temperature patterns that influence the winds. Acting in concert, the ocean and atmosphere are capable of music that neither can produce on its own. El Niño is an example of such music. An even broader range of climate fluctuations becomes possible once the interactions involve, not only the atmosphere and hydrosphere, but also the cryosphere (Earth's ice volumes) and biosphere. Scientists have only started to explore these aspects of Earth's natural variability.

This chapter concerns only one aspect of our planet's music, its weather. (The next two chapters deal with the ocean and with El Niño, respectively.) Here we address what is perhaps the most interesting question in atmospheric sciences, why is there weather? The source of energy for atmospheric motion, sunlight, varies in essentially the same manner from one day to the next. Why then are atmospheric conditions different from one day to the next?

Why There Is Weather

To explain weather, it is useful to explore the nature of sound. Sound involves motion, usually the vibrations of an object. Every object is elastic to some extent so that it vibrates when struck; the vibrations produce sound. We can produce a variety of sounds by tapping, with a spoon, various objects such as a table, a hand, a pot. In general, the vibrations die away quickly, and the sounds, if audible, are often cacophonous. If we happen to tap a bell, then we conclude that it is a very exceptional object. In response to a gentle tap, a bell emits a clear, beautiful sound that lingers for a surprisingly long time. Like a tuning fork, or any musical instrument for that matter, a bell is designed to have vibrations that are easily excited and that are relatively simple so that the sounds correspond, not to the cacophonous juxtaposition of a large number of musical notes, but to only a few harmonious notes. It is generally difficult to elicit a pure note from a solid object other than a musical instrument, but it is easy to do so in a fluid and therein lies the secret of our planet's music; its atmosphere and ocean are fluids. A fluid is as musical as a bell. Even a child can produce a pure note in a pond by merely dropping a pebble into the water. The perfectly concentric, expanding rings on the surface correspond to a pure, ringing oscillation that we can see but not hear.

Our planet's music is as spontaneous as that of a bell or other musical instrument and, for that reason, is known as its natural variability. A breeze that blows over the ocean, like a bow that strokes a violin string, readily elicits music in the form of waves. The audible sound of a violin is soothing when the pressure from the bow is gentle and becomes strident when the pressure is great. The ocean's music changes similarly from ripples, to choppy whitecaps, to foaming, lashing waves as the gentle breeze grows stiff, turns into a strong wind, and becomes a gale that whips the ocean into a frenzy. This transition, from innocent ripples to an angry, frenzied ocean requires, not winds that fluctuate more and more wildly, but steady winds that grow in intensity.

The transition to turbulence that characterizes waves on the surface of the ocean when the speed of the wind increases can be simulated in a small tank filled with two immiscible fluids, one denser than the other. (Many hardware and nature stores carry such wave tanks with the two fluids in different colors.) When the tank is tilted, the two fluids move in opposite directions. This relative motion—similar to that of a breeze of air (one fluid) over the ocean (the lower, denser fluid)—excites waves at the interface. Waves may develop even if there is only one fluid that flows with a speed that varies with height.

Waves readily appear in the region where the speed changes rapidly, the region of shear. If atmospheric winds vary with height in such a manner, and if clouds happen to be present where the shear is large, then the cloud pattern can resemble a train of waves as in figure 7.1. The waves are, of course, present in regions of wind shear even when there are no clouds; they can give airplanes a very bumpy ride.

The music of the atmosphere covers a spectrum of scales that range from those that are observable by watching the local sky, as in figure 7.1, to those with global dimensions. The latter are associated with meanders of the Jet Streams that circle the globe, one in each hemisphere. Sometimes those meanders are so energetic that loops fold back on themselves and create complex patterns that include cyclones, anticyclones (discussed in chap. 6) and various other phenomena associated with weather.

An apparatus that cleverly illustrates how meanders of the Jet Stream, and hence weather, depend on two critical parameters—the temperature difference between the equator and poles, and the rate of rotation of the Earth—is a doughnut-shaped ring, an annulus, filled with water. The ring represents the atmosphere and is placed on a rotating turntable. In the absence of any temperature differences, the annulus and water rotate as if they form a rigid body; the atmosphere does not move relative to the rotating Earth. Cooling the outside ring of the annulus (the pole)—by means of ice, for example—and keeping the inside ring (the equator) warm causes the water to move relative to the annulus, simulating the atmosphere's Jet Stream. The speed of the flow (the intensity of the Jet Stream) is proportional to the temperature difference between the equator and pole. Hence, as the temperature difference increases, the jet moves faster and faster. At low speeds, for small temperature differences, the flow is in perfect circles. Once the jet attains a certain critical speed, however, meanders appear spontaneously. At first they are modest undulations, but at sufficiently high speeds—for sufficiently large temperature differences between the equator and poles—the gentle undulations grow into the wild, chaotic oscillations shown in figure 7.2.

The results in figure 7.2 explain why Shelley's west wind is wild in autumn and winter rather than summer: the temperature difference between the equator and pole, and hence the intensity of the Jet Stream, is greater in winter than in summer. Summer is not merely warmer than winter; weather in summer is different from that in winter. In a city such as Chicago—and in midlatitudes in general—January, but not July, can be a month of rapid changes. In the course of a day, a jump in temperature from a pleasant 20°C to well below freezing is not uncommon in January but is practically unheard of in July.

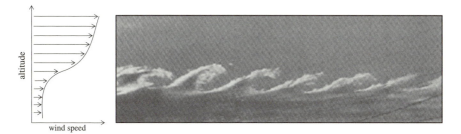

Figure 7.1 The wave patterns seen in the clouds are probably associated with winds that vary with height as shown in the sketch to the left.

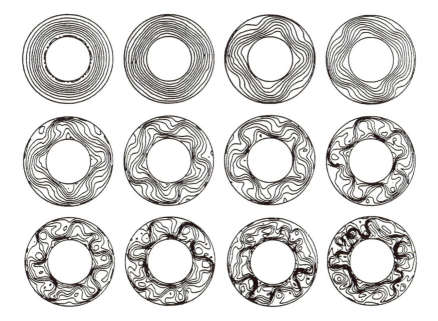

Figure 7.2 Flow patterns of a fluid in a rotating annulus when the inner and outer cylinders are at different temperatures. The lines correspond to isotherms at mid-depth. The transition from laminar to turbulent flow in this sequence of figures, starting at the top left, is a consequence of an increase in the rate of rotation from one figure to the next. The results are similar if the temperature difference is varied. From Buzyna et al. (1984).

Winter is often frenzied, whereas summer is relaxed, except for outbursts in the form of thunderstorms, usually late in the afternoon.

Temperature contrasts between the equator and poles, and between land and sea, set the atmosphere in motion and create a web of winds, the invisible threads of the climate tapestry described in chapter 6. Those threads, that range from gentle sea breezes to the majestic Jet Streams, are so taut that they vibrate spontaneously, thus producing weather, the music of our atmosphere. In the same way that a change in the tension of a violin string alters the note it produces, so a change in the speed of the winds alters the character of their vibrations. Such changes occur seasonally or, more generally, whenever Earth's climate changes. It follows that the character of weather depends on the climate. The opposite is also true; the day-to-day fluctuations in atmospheric conditions, the weather, can influence climate, the long-term average of those conditions. Consider a blustering storm in the heart of winter that makes us dream of warm, lush tropical islands. Such a storm actually contributes to pleasantly rather than unbearably hot conditions in the tropics, and to moderately rather than intolerably cold conditions in higher latitudes. A storm can effect a transport of heat from the equator to the poles. For example, the meander of the Jet Stream in figure 7.3 moves cold Arctic air toward the equator in the central part of the United States and also moves warmer air poleward along the eastern seaboard. The net effect of this flow of air is to cool the lower latitudes and to warm the higher latitudes (fig. 7.4). Although we may regard a winter storm as a nuisance, it increases the range of habitable latitudes considerably. Weather, because it is instrumental in transporting heat poleward, influences climate (fig. 7.4).

Experiments such as those in figure 7.2 can be used to explore various aspects of the complex interplay between climate and weather. For example, the apparatus permits studies relevant to the weather and climate of other planets because, in the apparatus, it is easy to change the rate of rotation to a value appropriate for another planet. Another powerful tool for such studies, one that can also be used to predict the weather, is a computer model of the atmosphere.

Models of the Atmosphere

The movements of weather patterns across the surface of Earth may appear to be chaotic but, in reality, are in strict accord with certain natural laws that can be used to predict the weather. Those laws, which identify invariants in our continually changing world—in es-

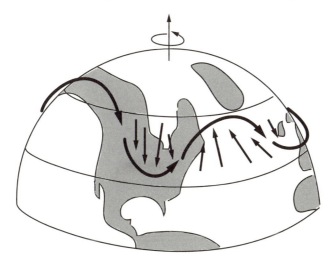

Figure 7.3 A schematic diagram of meanders of the Jet Stream that brings cold air to the central United States, warm air to the eastern seaboard, thus effecting a poleward transport of heat.

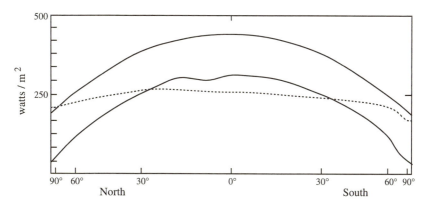

Figure 7.4 The upper solid curve is the solar energy incident on Earth at different latitudes. After reflection of sunlight, Earth's surface absorbs energy corresponding to the lower solid curve. The dotted curve shows the heat that Earth radiates to space. Earth is seen to absorb more heat than it radiates into space, in low latitudes. In high latitudes, the reverse is true. Both atmospheric winds and oceanic currents contribute to the poleward transport of heat implied by this result.

sence, they amount to Democritus' statement that "nothing can arise out of nothing, nothing can be reduced to nothing"—are very general and govern far more than the phenomena associated with weather; they govern all natural phenomena. The laws are therefore difficult to exploit unless we introduce simplifications that focus on the questions we wish to address and that filter out those that do not interest us. Such a procedure is known as constructing a model.

Let us start by addressing a relatively simple question, what is the globally averaged temperature of a planet that absorbs all its incident sunlight? A model suitable for this purpose regards each planet as a perfect, isolated sphere and invokes only the law that proclaims energy to be an invariant. That law demands that a planet radiates to space exactly as much energy as it absorbs in the form of sunlight. Given the incident energy—it can be determined by means of measurements—we can therefore infer how much energy a planet radiates and hence can calculate the temperature of the planet. The procedure amounts to that of an accurate accountant; by ensuring a balanced budget for energy, it determines the temperature of a planet. This information is minimal, a single temperature for the entire planet, but it is nonetheless useful because the difference between the calculated and actual surface temperature provides a measure of the greenhouse effect of the atmosphere (as explained in chap. 3).

A model that explicitly includes the greenhouse effect must take into account the atmosphere. The simplest model capable of doing this assumes that the atmosphere consists of layers, stacked on one another like pancakes. If the atmosphere is assumed to be static, then each layer must satisfy the law for the conservation of energy; each must have a balanced energy budget. Such a model explains the decrease in atmospheric temperatures with elevation. The results are qualitatively correct, but the calculated temperatures for Earth's surface, and the rate at which temperatures decrease, are much higher than those we measure. The problem is not with the bookkeeping—energy budgets in the model are balanced—but is in the degree of idealization: the absence of convection that redistributes heat vertically. Inclusion of this process yields results that explain why the climatic zones along the slopes of a mountain are similar to those we encounter when traveling from equator to pole: jungles, savannas, steppes, deciduous forests, pine forests, and tundra. Decreasing temperatures contribute to the changes in vegetation. The temperature difference between the equator and poles matches that between the base and the peak of a mountain at the equator, provided the peak reaches an altitude of 5000 m (15,000 ft). (With increasing elevation, the rate of decrease in temperature is approximately 10°C for every kilometer of altitude.)

The verdurous windward side and the arid lee of a mountain have different climatic zones, shown schematically in figure 7.5, which cannot be explained in terms of temperature variations only. Moisture and the direction of the winds are additional critical factors. Winds that approach a mountain carry air that is forced to rise. Because temperature and pressure decrease with elevation, the parcels of air expand. The expansion of air cools it—air that rushes from a bicycle tire feels cold for the same reason—so that the moisture in the rising air condenses into clouds that produce rain or snow. Once the winds have crossed the peak of the mountain and descend along the lee, the air parcels move into surroundings where the pressure is higher. The compression causes heating for the same reason that temperatures rise when air is pumped (or compressed) into a bicycle tire. The relative humidity of the air therefore decreases, as does the likelihood of rainfall, which is why the lee is usually sunny and arid.

For a climate model to reproduce the climatic zones in figure 7.5, it must divide the atmosphere, not simply into layers stacked on top of each other, but into boxes as shown. It is no longer sufficient to rely only on the law for the conservation of energy for each box; laws for the conservation of mass, moisture, and momentum (Newton's law of motion) all come into play. One of them amounts to the statement that nature abhors a vacuum. It is another way of saying that no climatic zone can continually lose or gain air. If the wind from the right carries air into one zone, then the wind must carry the same amount of air, as measured by its mass, out of that zone. The exchange of moisture is governed by a similar conservation law; moisture can neither be created nor destroyed in a zone, but it is possible for moisture to enter a zone in one form, as the gas water vapor, and, after condensation and the formation of clouds, to leave the zone in another form, as rain. In summary, a model of the atmosphere divides the atmosphere into climatic zones (boxes), each of which satisfies strict conservation laws which ensure balanced budgets in any exchange (of moisture, mass, etc.) between zones. Some of these budgets cannot be balanced independently of one another. If, in a certain box, high up on the mountain, water vapor condenses then that transformation is accompanied by the release of latent heat. The moisture budget therefore affects the energy budget, which, in turn, affects the momentum budget because the release of latent heat makes the air more buoyant so that it rises more readily. The bookkeeping in a climate model is in principle straightforward, but in practice it can become very compli-

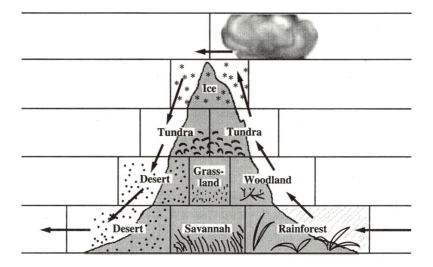

Figure 7.5 The climatic zones along the slopes of a mountain. The wind, indicated by arrows, blows from the right. A model divides the various climatic zones into simple geometrical boxes that facilitate calculations.

cated and involves an enormous number of calculations when the atmosphere is divided into a large number of climatic zones (fig. 7.6.)

If each climatic zone could be made infinitesimally small, then the methods of calculus become applicable and the various balanced budgets provide us with the equations that describe atmospheric motion. This means that a climate model that divides the atmosphere into a finite number of boxes provides an approximate solution to the equations that govern atmospheric motion. The accuracy of the solution, and hence of the simulation of Earth's weather and climate, increases as the number of boxes and balanced budgets increases. To cope with a large number of boxes and budgets requires an electronic computer. The most powerful computers available in the early 1990s can cope with models that divide the atmosphere into boxes, each of which covers approximately 100 km by 100 km with a dozen zones in the vertical. Such models still fail to resolve numerous important atmospheric phenomena, those that cover smaller areas, for example. By far the biggest challenge is to cope with clouds. Some of their properties depend on the size of their water droplets. Clever methods have to be devised to include such "subgrid scale" features in models. Because of the limitations of the computers, compromises are necessary. Different research groups proceed differently which is the reason for differences between models.

Figure 7.6 A computer model of the global atmosphere is composed of boxes that correspond to distinct climatic zones.

This description of models of the atmosphere focuses on their use in calculating equilibrium states. For example, the simplest model, which regards the entire Earth as one box, calculates an equilibrium temperature by assuming that the planet absorbs as much radiant heat as it emits. A generalization of that approach leads to a model that, by dividing the atmosphere into a huge number of boxes, simulates Earth's diverse climatic zones. For the model to succeed, it is necessary to specify the intensity and distribution of the incident solar radiation, the rate of rotation of the planet, the chemical composition of its atmosphere, and the properties (temperature, albedo, etc.) of the land, water, and ice surfaces beneath the atmosphere. Appropriate changes in these specified quantities permit simulations of climates on other planets or climates in Earth's past. The model is likely to reproduce weather, too, but the day-to-day variations in atmospheric conditions do not correspond to the weather on any particular day and are of interest only to the extent that they influence long-term averages. Thus, the model succeeds in reproducing, not the weather on a specific day in July, say, but the average conditions (temperature, rainfall, etc.) for the month of July.

A model of the atmosphere that balances budgets is also capable of predicting the weather, provided the definition of a balanced budget is generalized. Let us consider, for a moment, a bank account. If we deposit as much money as we withdraw each month, then the account is in a steady equilibrium state. Should we deposit more than we withdraw, then the account grows each month. It is possible to calculate at any time exactly how much money the account has, provided we know how much money it had initially. A model of the atmosphere, by balancing budgets in this generalized sense, can similarly calculate how conditions in every box change from one day to the next, provided an accurate description of the initial state of each box is available. Chapter 2 discusses how errors in the description of the initial state—failure to take into account a butterfly that flaps its wings, for example—amplify and cause predictions of subsequent states to become more and more inaccurate. That is why the success of weather predictions depends critically on accurate measurements of the state of the atmosphere.

Predicting the Weather

The value of meteorological data for the purpose of weather forecasts became increasingly clear during the second half of the nineteenth century because of incidents such as the destruction of a French fleet in the Black Sea in 1854. The storm that caused that damage caught the admirals by surprise, but it subsequently emerged that the storm could have been anticipated. Over the course of a few days, it had traveled eastward across southern Europe, leaving a trail of disaster in its wake. Word could have been sent to the Black Sea before disaster struck by using the telegraph, which had been invented in 1843. That invention was of enormous meteorological importance because it permitted an integrated network of observing stations. Many countries established meteorological services in the latter half of the nineteenth century and started to exchange data in "real time," within a matter of hours after the measurements had been taken. At first the data were surface measurements of pressure, temperature, and rainfall from land and sea. As explained in chapter 6, information about pressure readily provides information about winds; they tend to blow parallel to lines of constant pressure (isobars). Careful analyses of a sequence of pressure maps reveal that there are recurrent themes in the apparently chaotically changing weather patterns. This first became evident to a group of Norwegian meteorologists early in the twentieth century. They started to track the motion of different masses of air and, in a choice of terms that probably reflected the influence of World War I, designated the lines along which warm and cold masses of air clash as "fronts." A sequence of maps reveals the direction in which fronts and other weather patterns are propagating and hence can serve as the basis for a weather forecast.

The flow of air is somewhat simpler aloft than it is near the surface, where the details of local topography can complicate matters enormously. Furthermore, changes in the meanders of the Jet Stream often precede developments near the surface, which is why measurements in the upper atmosphere that complement those at the surface are particularly valuable. They can be made by means of a relatively cheap instrument, the expendable balloon, whose movements are tracked from the ground. A balloon measures the temperature, pressure, humidity, and other variables and relays the information back to the surface. The number of atmospheric soundings with balloons increased considerably during World War II. Today, some 700 are released worldwide twice a day, at noon and midnight Greenwich Mean Time. (They are frequently mistaken for flying saucers.) A revolution in our ability to observe the atmosphere occurred in the 1960s with the advent of meteorological satellites. At present, there are several polar-orbiting satellites, plus four geostationary ones over the equator, to ensure continual coverage of the globe. The vast stream of data, from the satellites and the huge network of stations, flows from national centers through 21 Regional Meteorological Centers to 3 World Meteorological Centers (Washington, Moscow, and Melbourne), where data are collated and charts are generated. The information is available to all nations. If they have the appropriate receiving systems, they also have direct access to the automatic picture transmissions from American weather satellites. Today virtually every country participates in World Weather Watch, organized by the World Meteorological Organization in Geneva. Quarrelsome mankind is more truly united in this activity than in any other.

Countries eagerly participate in World Weather Watch because weather is a global phenomenon; everybody needs everybody else's information. To cope with that information is a major undertaking that requires mammoth supercomputers. Of what use is the computer if there are errors in the measurements? (One of the most common errors in the reports from ships is in their position; by typing in the wrong number for longitude, they sometimes give locations that are on land, in the Sahara desert, for example.) Lack of data is another problem that seriously affects the preparation of weather maps. Vast as the flow of meteorological data may seem, coverage is sparse over the oceans, especially in the southern hemisphere, which is mostly ocean. How can an accurate weather map be drawn without data? It is possible to mitigate these problems by making use of a computer model that simulates the atmosphere. Let us for the moment assume that such a tool exists and that it is perfect. Starting from a perfect description of atmospheric conditions at a certain time, such a model would reproduce the subsequent evolution of weather patterns with complete accuracy and there would be no further need for any measurements. The models, however, are far from perfect. Our other source of information, the measurements, also has inaccuracies. By combining these sources, their flaws can be minimized. Such an approach provides the most reliable description of meteorological conditions. The problem is similar to that encountered by experimental scientists who wish to determine the slope of a straight line that expresses the relation between certain variables, such as temperature and pressure, for example. In principle, they need only two points on a graph—that is all that is required to draw a straight line. However, the measurements that determine the two points have errors. The experimentalists therefore make a larger number of measurements in order to have a great many points on the graph. It is then a matter of deciding which straight line fits the data best. The meteorological problem is in principle the same, in practice much more complex. It requires a compromise between the model and the measurements. This procedure can be done objectively and has advanced to such a level that scientists at the World Meteorological Centers can identify the faulty instruments at various stations around the globe that consistently give inaccurate measurements. The weather maps that appear daily in newspapers and on television are products of an enormous global network of meteorological measurements plus sophisticated computer models of the atmosphere.

The Storm of April Fool's Day 1997: A Case Study

Everybody in New England talked about the weather on April Fool's Day of 1997. On that Tuesday morning, everything was covered with a thick blanket of snow even though the preceding Easter weekend had been warm and sunny. After the storm, spring returned quickly; it was back by the end of the week. What had caused the dramatic interlude? And how were the weather forecasters able to predict it so accurately?

To understand how storms develop, and to appreciate what is involved in making weather forecasts, it is a useful exercise to keep a weather log (for a while, anyway). The first three presidents of the

United States, George Washington, James Madison, and Thomas Jefferson, all kept such logs that included regular thermometer readings at their homes; Jefferson did so for 50 years, starting in 1777. A weather log of local conditions usually records temperature, pressure, extent of cloudiness, and type of precipitation, if any. Such information about changes in local atmospheric conditions facilitates identification of recurrent patterns. For example, a decrease in pressure often indicates that foul weather is imminent. Today an amateur meteorologist, far more easily than his counterpart of the eighteenth century, can anticipate evolving patterns by turning to the World Wide Web, which has a wealth of information about atmospheric conditions.

In Princeton, New Jersey, Easter Sunday 1997 started out mild and sunny and continued that way into the afternoon. The only local indication of an impending storm was a gradual drop in pressure. Weather buffs who also recorded dew point temperature took note that the difference between the actual temperature and the dew point was slight. Those who turned to the World Wide Web for further information learned of similar conditions over a large part of the eastern seaboard. That information, if plotted on a map as has been done in figure 7.7a, shows, over the northeastern seaboard on Sunday evening, a relatively weak low-pressure system associated with modest winds and some precipitation in the form of rain (except where temperatures were sufficiently cold for snow). Figure 7.7b shows how that weak cyclone, over the next 24 hours, exploded into an intense storm with fierce winds howling around its center of very low pressure. (In New England, which was buried in a few feet of snow, the storm was referred to as a north-easter because the winds in that region were from that direction.) How was it possible, on Sunday evening, to anticipate the developments of the next 24 hours?

The growth of any storm requires complementary conditions at the surface and aloft. A low-pressure system, as explained in chapter 6, is associated with anticlockwise winds that, at the surface, spiral inward. As a consequence, the air is obliged to rise over the center of low pressure, causing the temperature of the air to fall, moisture to condense, and clouds to form. If conditions in the upper troposphere should remain unaffected, then more and more air accumulates over the center of low pressure. The pressure will therefore not remain low, and the storm will decay. For the storm to amplify, the convergent flow near the surface must be complemented by divergent flow aloft so that the pressure at the surface can remain low.

The storm of April Fool's Day 1997 intensified rapidly because of favorable conditions aloft. This could have been anticipated on Sun-

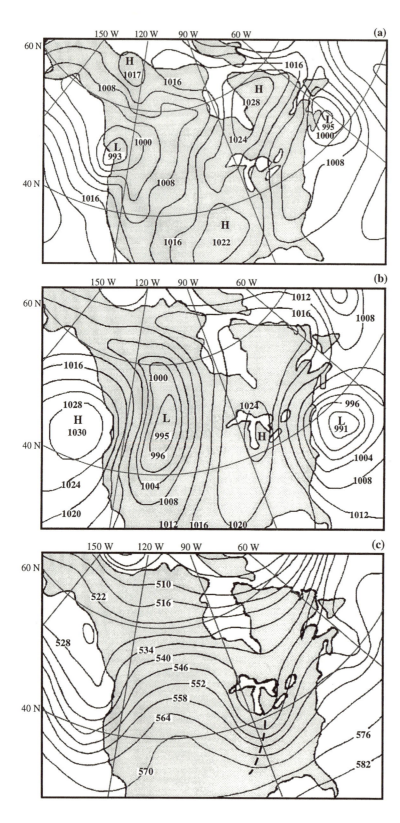

Figure 7.7
Maps of surface pressure (in mb) on the evenings of March 30 (a) and 31 (b), 1997. Note how the low-pressure zone over New England intensified considerably over a period of 24 hours. The reason is evident in (c), the map of the height of the 500-mb surface on March 30. The key feature is the trough, the dashed line, to the west of the surface low-pressure zone. Based on charts provided by the Atmospheric Environment Service of Canada.

day evening by inspecting a map of conditions aloft, (e.g., 7.7c). As is often the case, a meander of the jet stream was at the heart of the matter. Its southward loop around a center of low pressure caused a trough in the streamlines. Upstream from the trough (the dashed line in fig. 7.7c), the streamlines squeezed together so that the flow in that region was convergent. Some of the air was therefore forced downward, toward the surface. Downstream from the trough the opposite happened: the flow was divergent, and air was sucked up from the lower layers of the atmosphere. This divergence immediately above the center of low pressure at the surface intensified that center, causing stronger convergent flow at the surface. The elevation of relatively warm, moist air and its replacement by colder air released potential energy that became the kinetic energy of strong winds.

The amplification of the storm of April 1, 1997, depended critically on the centers of low pressure, aloft and at the surface, being displaced relative to one another; the center aloft must have been to the west of the one at the surface. That displacement ensured complementary conditions at the surface and aloft. A second factor that contributed to the growth of the storm was its location over the eastern coast of the United States. The winds were able to acquire moisture and heat from a vast reservoir, the relatively warm ocean. In that respect the storm resembled a hurricane.

In the northeastern United States, the storm of April 1, 1997, ended a relatively mild winter. The preceding winter, that of 1995–6 had been far more severe. (It had several storms similar to the one described above except that a few developed much farther south and traveled northeastward along the coast, intensifying along the way and affecting a much larger area. Atmospheric scientists were as accurate in predicting those storms as they were in forecasting the storm of April 1, 1997. However, they were not able to predict that the winter of 1995 to 1996 would be severe, or that the following one would be mild. Why was the one winter different from the other? What is required to anticipate, not a specific storm a few days in advance, but the mildness or severity of an entire season?

On a chilly day in January, it is possible to forecast accurately that a typical day in July, 6 months later, will be much warmer. An atmospheric model that predicts the weather can match that feat. It is capable, in January, of predicting conditions in July, provided the change in the distribution of sunshine, and hence in the temperature patterns at Earth's surface, are specified. That accomplishment merely matches a forecast based strictly on data that describe past seasons. For a model to do better, it must exploit the fact that a change in conditions at Earth's surface, from one winter to the next, contributes to the difference between one season and the next. Much of the Earth's surface is covered by water, and a change in sea surface temperatures can have a profound effect on atmospheric conditions. Thus, to understand why some winters are mild and some severe, we next turn our attention to the oceans.

8

THE OCEAN IN MOTION

HALLENGER, Discoverer, Atlantis, Meteor, Calypso . . . The evocative names of oceanographic research vessels enhance the aura of romance of oceanographic expeditions. The sponsors of those expeditions, however, usually have practical goals in mind. When the British corvette HMS *Challenger* circumnavigated the globe between 1872 and 1876—the science of oceanography dates from that period—her goals included the collection of bathymetric data needed to lay telegraph cables across the ocean floor. When oceanography flourished during the Cold War, because of generous support from governments, the justification was primarily the military importance of the oceans. (Submarines are difficult to detect because seawater conducts electricity, thus ruling out the use of tools such as radar.) The end of the Cold War, and the consequent reduction in military budgets, is worrisome to oceanographers. Who will now support their research? Is there still a role for those who wish to explore the oceans with "a tall ship, and a star to steer her by"?

A physicist recently explained to a politician that his research would not contribute to the defense of the country but would make the country worth defending. Although some scientists share this belief that their research is an end in itself—it may give their country international prestige or may answer fundamental questions that we have a deep need to have answered, questions about the origin of the universe, for example—most argue that their results, in the long run, will prove useful. There are numerous examples of abstruse scientific results that have subsequently been translated into technological marvels. It nonetheless is an advantage to be useful in a direct and obvious way. Meteorologists benefit from the public's need to know of impending storms and of the whereabouts of hurricanes. They meet that need by forecasting the weather on a routine basis, day after day. This obligation may seem onerous but, in many respects, is beneficial. Operational weather prediction, which blossomed in the 1950s shortly after the invention of the electronic computer, requires the close collaboration of specialists with diverse interests: computer scientists, meteorologists, applied mathematicians, engineers who develop instruments that measure the atmospheric conditions, and so on. Weather forecasting not only integrates the activities of these different groups, thus creating a cohesive community, but also justifies a global network of observational stations and supercomputers. These riches are envied by oceanographers who are aware that all the oceanographic data ever collected amount to less data than meteorologists routinely gather within a matter of days. There is now a risk that the acquisition of even the slender amount of oceanographic data will slow down because one of the principal sponsors of this science, the military, is having its budgets reduced. Oceanographers and a great many other scientists find that, in the aftermath of the Cold War, their potential sponsors insist on results that will benefit society in a direct and obvious manner.

Oceanographers are fortunate in that their services are essential for a variety of activities of enormous public interest. One is the prediction of climatic changes. Because it exchanges large amounts of heat, moisture, and gases such as carbon dioxide with the atmosphere, the ocean influences Earth's climate in a variety of ways. For example, fluctuations in those exchanges play a role in determining whether next winter will be exceptionally cold, whether the following summer will be unusually wet, or whether El Niño will occur. Because of the need to monitor the oceans, scientists have initiated activities that amount to the counterpart of operational weather predictions, thereby transforming the science of oceanography. The new era promises to be as beneficial to oceanography as operational weather forecasting is to atmospheric sciences. It is justifying a growing network of automated oceanic measurement and is integrating the efforts of scientists with diverse interests ranging from the physics, chemistry, and biology of the oceans to the design of computers and the development of mathematical techniques to solve oceanographic problems. Some traditionalists have qualms about this change. Although oceanography will inevitably lose some of its romance in an era when navigators use an artificial satellite rather than "a star to steer her by," that loss is far outweighed by the obvious benefits.

The ocean is a mere film of salty water—4 km deep, on the average—on the surface of our globe, which has a diameter exceeding 12,000 km. The most remarkable aspect of that film is the stark contrast between the shallow, warm upper layer, some 100 m deep, where light is abundant and where most marine life can be found, and the cold, dark, deep abyss. Figure 8.1 depicts the thermal contrast and also shows that, whereas the deep ocean has plentiful supplies of nutrients and of carbon dioxide, the surface layer does not. Because a

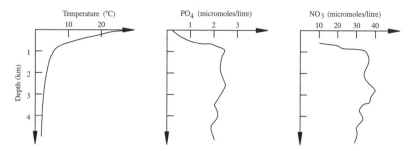

Figure 8.1 Typical vertical profile of temperature, phosphate, and nitrate in the ocean. Note the contrast between the surface layers and the deep ocean.

change in these vertical contrasts can profoundly affect Earth's climate, we need to explore the reasons for the contrasts.

The inability of sunlight to penetrate more than tens of meters below the ocean surface is the principal reason for the differences between the surface and deep layers of the ocean. The effectiveness with which seawater absorbs sunlight causes the surface layers to be relatively warm, the deeper layers cold. There is light only near the ocean surface, in the euphotic zone, so that oceanic plants, which require light for photosynthesis, must be able to float. Most are microscopic in size and are known as phytoplankton, literally "plants that wander." They are so small that a beaker of seawater can contain millions of invisibly small algae, bacteria, and protozoa. These numbers may sound impressive but in reality are not; otherwise the color of the ocean would be green. (Ponds with an abundance of fertilizers turn green because of the plants that grow in them.) The relative lack of plants near the ocean surface—much of the ocean is effectively a desert—is attributable to the low concentration of nutrients near the surface. In the deep ocean, nutrients are far more plentiful, as is evident in figure 8.1. The plants, the zooplankton that graze them, and other biota in the food chain absorb much of the nutrients and carbon dioxide that are available near the surface. When they die, they sink into the abyssal ocean, decompose, and break down into their constituent chemicals. The biota in effect pump carbon dioxide and nutrients from the surface layer into the deep ocean and contribute to the vertical contrast shown in figure 8.1. In principle, oceanic trees with deep roots should be able to take advantage of the nutrients at depth. Such trees have not evolved, probably because of the intense oceanic currents that are confined to the upper ocean (see Colinvaux 1978).

The warm upper ocean basks in sunlight, while the pitch dark abyss remains freezing cold. (The region of strong temperature gradients that separates the warm upper ocean from the cold water at depth is known as the thermocline.) From figure 8.2, which shows how temperatures vary with depth along a meridian in the Pacific, it is clear that the layer of warm surface water is so shallow that the average temperature of a column of seawater, even at the equator, is close to the freezing point. Unlike the atmosphere, which is heated efficiently from below so that convection readily redistributes heat vertically, the ocean is heated inefficiently from above. The warmest water remains at the surface, and the stable stratification inhibits vertical movements of water parcels. That the thermal contrast between the surface layers and the deep ocean has persisted for millennia poses a puzzle. Why has heat not diffused downward and increased temperature at depth? (If that should happen, then Earth's climate would change radically because the atmospheric concentration of carbon dioxide would increase considerably.) The capacity of seawater to hold carbon dioxide increases as the temperature of the water decreases, which is one reason why the concentration of carbon dioxide is greater in the deep ocean than in the surface layers. If the temperature of the deep waters should increase, then carbon dioxide would bubble out of the ocean. This has not happened—the thermocline remains shallow and the deep ocean cold—because of the motion of the ocean.

Oceanic currents are driven primarily by forces at the ocean surface: those associated with the exchange of heat and moisture between the ocean and atmosphere, and the winds. It is therefore not surprising that practically all the intense currents are in the surface layers, in and above the thermocline. The prevailing winds, which are in different directions in different latitudes, tend to drive the surface flow westward in the tropics and eastward in the subtropics. The Coriolis force deflects water parcels to their right (in the northern hemisphere), thus causing the flow to have a poleward component in the tropics, an equatorward component in the subtropics (see the arrows in fig. 8.2). These surface flows, from opposite directions, meet in the neighborhood of 30°N and 30°S where the water sinks. Some of it returns to low latitudes in subsurface layers with the same temperature, close to 18°C. At the equator, the water wells up into the surface layer, to be warmed by the atmosphere before returning to the subtropics. The remainder of the water that sinks near 30°N and 30°S participates in subtropical gyres and travel poleward in currents such as the Gulf Stream and Kuroshio, before returning to the region of sinking. These wind-driven circulations that effect exchanges between the tropics and extratropics are relatively shallow. The maintenance of

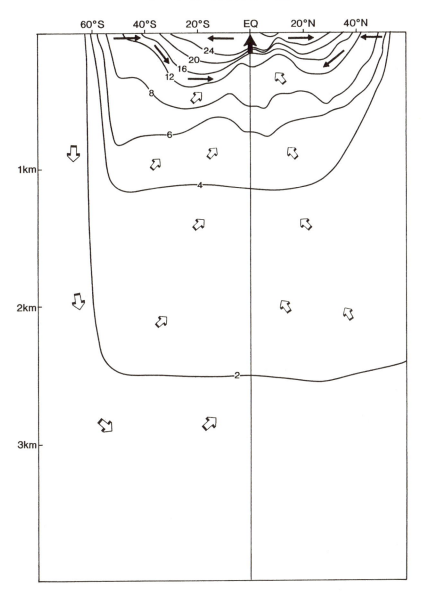

Figure 8.2 Isotherms (°C) below the ocean surface, in a vertical plane that runs north-south in the Pacific Ocean. The thermocline, the region of large temperature gradients that separates the warm surface layers from the cold water at depth, is particularly sharp and shallow near the equator. The oceanic circulations that maintain the thermocline are shown schematically by means of solid arrows, for the shallow wind-driven circulation, and open arrows for the deeper thermohaline circulation.

the thermocline depends not only on them but also on another circulation, with a much larger latitudinal and vertical scope. It is known as the *thermohaline circulation* and involves not only the winds but also depends critically on the salt of the sea.

The Thermohaline Circulation

Although it accounts for only 3% approximately of the ocean by weight, salt is of enormous importance because it affects the density of seawater. In the absence of salt, warm water, because it is less dense than colder water, will always float on the colder water. If, however, the density of the warm water were to increase because of the addition of salt, then the surface waters would sink in spite of being warmer than the deeper water. This happens most readily in high latitudes where the temperature difference between the upper and lower ocean is small and an increase in the salinity of the surface water causes it to sink. Two naturally occurring processes can induce such oceanic convection. One is evaporation, which removes pure water molecules, free of salt, from the ocean surface, leaving behind seawater with increased salinity. Extremely cold, dry winter air that moves from a continent onto the relatively warm ocean effects such an increase in the density of seawater. The air heats up over the ocean so that evaporation from the ocean increases, thus creating cold, saline water that sinks.

Another process that increases the salinity of the upper ocean is the formation of ice. The ice that forms when water freezes is free of salt. (Recall that Saudi Arabia, a few years ago, explored the possibility of importing icebergs from Antarctica as an alternative to local desalination of seawater). The cold, salty water below the sea ice is so dense that it sinks.

The sinking of cold, saline water in high latitudes necessitates a poleward flow of surface waters, an equatorward flow at depth, and rising motion to close the thermohaline circulation. This circulation, shown schematically in figure 8.2, is asymmetrical relative to the equator because the continents are arranged asymmetrically. The three principal oceans, the Indian, Atlantic, and Pacific Oceans, are connected only in the southern hemisphere, by the Antarctic Circumpolar Current. Off the southern flank of this current, near Antarctica, temperatures are so low that icebergs form, thus creating very salty sea water that sinks. In the northern hemisphere, the sinking of saline water occurs only in the Atlantic. The Pacific is insufficiently cold, and its water is insufficiently saline, for the formation of dense bottom water. Approximately half of the water in the deep ocean has its origin in the northern Atlantic. From there it flows southward, across the equator, to join the Antarctic Circumpolar Current. That current, in turn, loses cold, dense water to the Indian and then to the Pacific Ocean as shown schematically in figure 8.3. If we date a water parcel from the time that it leaves the surface and sinks into the deep ocean then the youngest water is in the deep northern Atlantic, and the oldest water is in the deep northern Pacific, where its age is estimated to be 1000 years. Corroborating evidence for the validity of this picture comes from the distribution of various chemicals. The concentration of nutrients, for example, is at a minimum in the deep northern Atlantic because the water there has just arrived from the nutrient-depleted surface layers. On its subsequent journey, the water is then steadily enriched by sinking organic material so that the most nutritious waters are those of the deep northern Pacific. These are also some of the most unpolluted waters on Earth. Man deposits contaminants directly into the ocean and also indirectly by releasing into the atmosphere gases that subsequently enter the ocean, such as CFCs and the chemicals tritium and C^{14}, which were released through atmospheric atomic bomb tests in the 1950s and 1960s. Oceanic measurements indicate that these chemicals entered the deep ocean in the northern Atlantic and are gradually moving southward. When last seen, they were crossing the equator (in the Atlantic). They should make their appearance in the abyssal northern Pacific a few centuries hence.

The water that sinks into the deep ocean is dense because of its low temperature and high salinity. If the salinity is high enough, the water sinks even if its temperature increases slightly. Thus, the atmospheric manifestation of global warming could be deferred should the oceans absorb the heat and store it in its deep layers. This may in fact be happening and could be one reason why global warming is so difficult to detect. The heat capacity of water is so high—it requires a relatively large amount of heat to raise the temperature of water—that even a very slight increase in the temperature of the deep ocean could represent a considerable sink of heat for the atmosphere. For this sink to operate, the increase in oceanic surface temperature must be gradual. Too large an increase will make the surface waters too buoyant and will inhibit formation of deep water. In such a case, the rate of global warming of the atmosphere will increase.

Surface water is insufficiently cold and saline to sink in the northern Pacific, but it is sufficiently dense to sink in the northern Atlantic.

Figure 8.3 Cold, saline water sinks primarily in the northern Atlantic and around Antarctica and then spreads to fill the deep oceans as shown schematically in this sketch. This water gradually rises to the surface to participate in the shallow, wind-driven circulation shown in more detail in figure 8.6.

One possible reason is the very salty water that flows into the Atlantic from the Mediterranean Sea (where evaporation exceeds precipitation). Another, probably more important reason is the self-sustaining nature of the thermohaline circulation, which can be halted by a decrease in the salinity of the surface waters. Just before they reach the region of sinking in the northern Atlantic, water parcels travel through a region of heavy rainfall (see fig. 8.4). This freshening of the water decreases its density so that a parcel must travel swiftly across the region of high precipitation if it is to be dense enough to sink at the end of its journey. The thermohaline circulation causes parcels to move rapidly across that zone of heavy rainfall in the Atlantic. In the absence of that circulation, a parcel lingers and is likely to become so fresh that its density will be too low for the parcel to sink. This means that the thermohaline circulation will be difficult to revive if it ever comes to a halt; on the other hand, it tends to sustain itself by rapidly moving water parcels out of the zone of precipitation.

Theoretical investigations of thermohaline circulations without sinking in the northern Atlantic indicate that the current state of affairs can, in principle, be replaced by three other possibilities. Calculations with an idealized model of the ocean, in which the Atlantic and

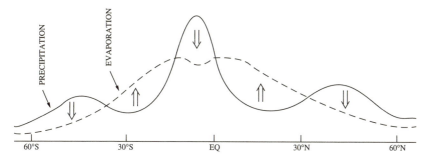

Figure 8.4 The latitudinal distribution of evaporation and precipitation. Surface waters of the ocean are saline where evaporation exceeds precipitation.

Pacific are linked by an Antarctic Circumpolar Current, as in figure 8.5, show that, depending on initial conditions, sinking in either, both, or neither the northern Atlantic and Pacific Oceans are all possibilities.

The sensitivity of the present thermohaline circulation to perturbations—what is required for it to switch off suddenly or change to one

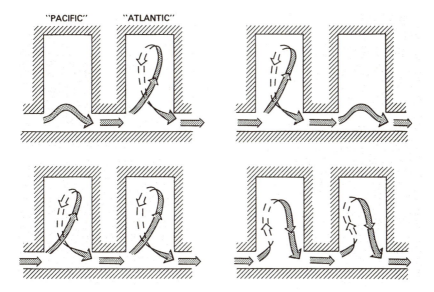

Figure 8.5 A schematic diagram of four possible climate states obtained in an ocean model. After Marotzky and Willebrand (1991).

of the other possibilities shown in figure 8.5—is unknown. Studies of paleoclimates could shed light on this important matter. The global warming since the last glacial maximum, some 18,000 years ago, has not been smooth and steady; prolonged cold spells interrupted the gradual warming (see chap. 11). Could changes in the thermohaline circulation have accompanied those climate fluctuations?

Dense water that sinks in high southern latitudes and in the northern Atlantic, spreads across the ocean floor and keeps the deep ocean cold. At this time, it is not known exactly where this water rises back to the surface to join the wind-driven currents discussed next.

The Wind-Driven Surface Currents

The wind-driven currents that carry water to the sinking regions are all confined to the upper ocean except for the Antarctic Circumpolar Current, which extends to considerable depths, practically to the ocean floor. There are two reasons for this exception: the weak vertical stratification of the ocean in high latitudes, which permits the wind-forcing to penetrate to great depths, and the enormous fetch of the winds around the Antarctic continent. There, and nowhere else, water parcels can circle the globe. Only the Antarctic Circumpolar Current links the three major ocean basins, the Atlantic, Indian, and Pacific Oceans.

Away from the high southern latitudes, water parcels tend to move in gigantic gyres driven by the torque exerted on the ocean by easterly winds in the tropics, westerlies further poleward. These gyres have a fascinating asymmetry because of the rotation and sphericity of Earth. Instead of a relatively uniform circular motion with equatorward flow in the eastern half of the basin and poleward flow in the western half, the poleward flows are concentrated in very narrow, swift jets that hug the western boundaries of the ocean basins before venturing offshore. The swiftness of these rivers of warm water off the eastern coasts of continents, the Gulf Stream in the Atlantic and the Kuroshio in the Pacific, for example, contrast sharply with the much broader, slower southward drift of cold water off the western coasts of continents. The equatorward California and Canary currents, for example, are far slower than the Gulf Stream. Figure 8.6 shows the major oceanic currents.

Direct measurements of oceanic currents—especially from a ship that drifts with the winds and currents, and whose location is not known precisely—are very difficult. It is, however, possible to infer

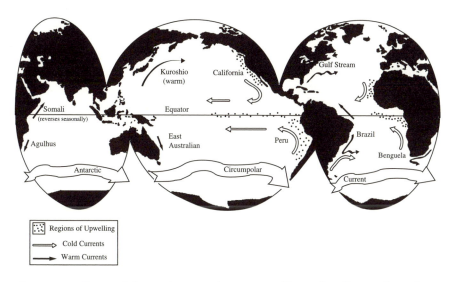

Figure 8.6   Some of the major oceanic currents. Those that flow poleward carry warm water; those that flow equatorward generally transport cold water. The Antarctic Circumpolar Current connects the three major ocean basins and also shields Antarctica from poleward-flowing currents that carry warm water. The dots indicate regions of intense upwelling that are rich in nutrients and hence are rich fishing grounds.

the currents from measurements of temperature and salinity. The trick is to calculate, from those measurements, the elevation of the sea surface. (The ocean surface is not level!) Water expands when it is heated so that an increase in the temperature of a column of water in the ocean causes the sea level to rise. Hence, once it is known how temperature (and salinity) vary with depth at various locations, it is possible to calculate how the height of the ocean surface changes from one location to the next. Recently, satellites have started to provide measurements of variations in the height of the sea surface. The results indicate that across currents such as the Antarctic Circumpolar Current, the Gulf Stream, and the Kuroshio, the height of the ocean surface changes by approximately a meter. Because of gravity, water will tend to flow from the elevated to depressed regions. However, a water parcel that starts to flow "downhill" comes under the influence of the Coriolis force (because of the rotation of Earth) and is deflected toward its right (in the northern hemisphere). This deflection continues until the parcel is flowing along lines of constant elevation, which is why the directions of currents usually coincide with lines of constant elevation. (The arguments are identical to those in chapter 6 in connection with atmospheric motion along rather than across isobars.)

Sea level on the Pacific side of central America is higher than on the Atlantic side. Water does not rush eastward through the Panama Canal, however, because the canal has two parts that are connected to a freshwater lake that is higher than either the Pacific or the Atlantic Ocean. If there were a canal that permitted water to flow directly from the Pacific to the Atlantic across Central America, then the current through that canal would be swift.

The ocean surface is higher in the western than the eastern equatorial Pacific. Far from the equator, the Coriolis force prevents the water from simply flowing "downhill." At the equator, however, the Coriolis force vanishes, and the water does flow "downhill," from west to east, in an intense, narrow current that is confined to the immediate neighborhood of the equator. The winds that prevail there, the Trades, are in the opposite direction, westward, and drive the surface flow in that direction. The eastward current that is flowing "downhill" is therefore subsurface. Biologists fishing for tuna on the equator in the central Pacific in 1953 first became aware of this current when they noticed that, although their ship drifted westward with the wind and the surface currents, the long fishing lines moved eastward. Subsequent measurements confirmed the remarkable current, known as the Equatorial Undercurrent. In transport and speed it is comparable to the Gulf Stream. It flows just below the surface—the core is at a depth of 100 m—all along the equator, across the entire width of the Pacific (see fig. 8.7). Its width is a mere 200 km so that the current in effect is a ribbon, nearly 15,000 km in length, that marks the location of the equator.

When a ship crosses the equator, the occasion usually calls for a festive, amusing ceremony to initiate those who are doing this for the first time. The humor stems from the pretense that the line that is being crossed is real; everybody knows of course that it is imaginary. But actually the line is real! Along it lies not only the Equatorial Undercurrent but also a narrow tongue of cold surface waters with very high chlorophyll values. This tongue, which is clearly visible in satellite photographs when the easterly trade winds are strong, appears because those winds cause a parting of the surface waters. Water parcels that are driven westward by the winds experience a Coriolis force that deflects them northward if they are slightly north of the equator, southward if they are to the south. This divergence of the surface waters from the equator causes the upwelling of cold, nutri-

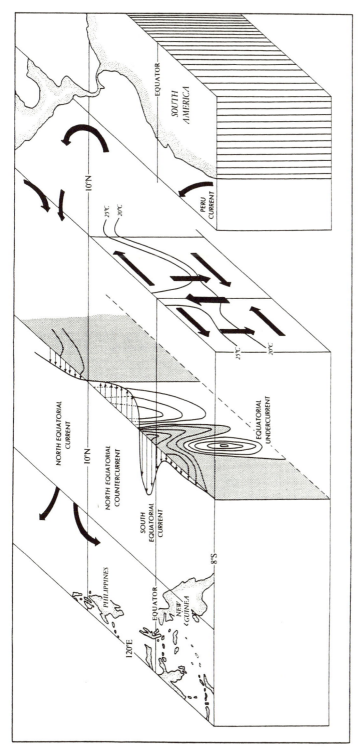

Figure 8.7 The alternating bands of eastward and westward flowing tropical currents that are connected by circulations in the north-south plane.

ent-rich water from below, making the equatorial zone one of the bio-
logically most productive regions of the oceans.

Oceanic upwelling also occurs along the California coast and
causes the surface waters there to be cold and rich in nutrients. The
prevailing winds have an equatorward component parallel to the
coast. At first they drive the surface waters southward, but then the
Coriolis force comes into play and deflects the waters offshore. As a
result, cold nutrient-rich water comes to the surface along the coast.
Figure 8.6 shows the various coastal zones where the prevailing
winds induce upwelling. All are rich fishing grounds.

Intense oceanic upwelling occurs primarily along parts of the equator and certain coasts. Downwelling, too, is confined to small regions.
The sinking of dense, saltwater is limited to high northern latitudes in
the Atlantic and to the neighborhood of Antarctica. In addition, there
are subduction zones (where surface waters sink) in the subtropics.
The main reasons for the latter regions are the easterly trade winds in
the tropics and the westerly winds in the subtropics. Because of the
Coriolis force, the surface currents induced by those winds have a
poleward component in low latitudes and an equatorward component in higher latitudes. These waters move in opposite directions
and meet in the neighborhood of 30°N and 30°S, where sinking occurs. This is an idealized picture; in reality, subduction is associated
with intense winter storms that mix surface waters to the depth of the
thermocline, which is said to be ventilated. The fluid parcels at depth
continue to travel in the thermocline and in effect contribute to its
maintenance. Some participate in the subtropical, wind-driven gyres,
some in the complex system of wind-driven equatorial currents.

A parcel can travel across the globe by means of the various currents in figure 8.6. Consider a parcel that circles the North Pacific in
the subtropical gyre of that ocean. The journey involves sinking motion off the coast of California and then westward motion. Then
comes the option of either returning poleward in the Kuroshio current
along the coast of Japan or joining the tropical circulation by proceed-
ing southward in the thermocline. If the parcel reaches the equator, it
will start to rise and join the westward surface flow. Soon it finds
itself drifting past exotic isles—Java, Sumatra, Borneo—and into the
Indian Ocean. There it may have to dawdle, waiting for the right
monsoon winds to carry it toward Africa and into the Agulhas current for the trek past Durban to the Cape of Good Hope.

There is an alternative route to the same destination. While in the
tropical Pacific, the water parcel could journey back and forth across
that vast ocean in its complex maze of east-west flowing tropical currents and undercurrents (fig. 8.7) until it finds itself in the southern

hemisphere. Slow southward drift then carries it into the Antarctic Circumpolar Current. It could circle Antarctica a few times but sooner or later, after emerging from the Drake Passage, the water parcel will peel off with the northward Malvinas currents and will head for a rendezvous with the warm, southward Brazilian current. The two embrace near Buenos Aires and tango off toward Cape Town. Satellite photographs clearly show how intertwined they are.

Once it finds itself off Cape Town in the South Atlantic, our water parcel slowly winds its way northward, first in the Benguela current along southwest Africa, then across the Atlantic toward Brazil where it joins the coastal current that carries it across the equator, into the Gulf of Mexico, out through the straits between Florida and Cuba, and finally northward in the Gulf Stream.

During this odyssey, the parcel experiences extremes of temperatures and salinity. It is fresh where it rains and salty in arid regions where evaporation is high. At the end of its journey, when it is depleted of nutrients and is cold and salty, it sinks into the pitch dark abyss and starts a long, slow journey across the ocean floor. During a journey of some thousand years it could visit each of the ocean basins before rising in the Antarctic Circumpolar Current to rejoin the wind-driven surface currents. Similar travels by an unending stream of seawater parcels—at a glacial pace in the deep oceans, more rapidly but more circuitously in the upper ocean—maintain the contrast between the surface layers and the abyssal ocean, the contrast that is of vital importance to Earth's climate.

Oceanic Weather

When he was postmaster general of the Colonies, Benjamin Franklin, prepared a map of the Gulf Stream after he learned from his cousin Timothy Folger why mail packets required 2 weeks less to journey from New England to England than to make the reverse trip.

> We are well acquainted with the stream because in our pursuit of whales, which keep to the side of it but are not met within it, we run along the side and frequently cross it to change our side, and in crossing it have sometimes met and spoke with those packets who were in the middle of it stemming it. We have informed them that they were stemming a current that was against them to the value of three miles an hour and advised them to cross it, but they were too wise to be counseled by simple American fishermen.

Presumably Franklin's chart of the Gulf Stream speeded up the mail service. Regular satellite photographs enable us to do better today; they provide us with information about the ever changing path of the Gulf Stream. It is now even possible to predict what the path of currents such as the Gulf Stream and Kuroshio will be a week or two hence—by means of computer models of the ocean, the counterparts of the models that are used to predict weather.

The ocean has two types of variability: one forced by the fluctuations of the winds that drive the ocean, and one that would be present even if the winds were perfectly steady. The most obvious example of the latter, natural variability of the ocean—its weather or music—are the waves that readily appear at the ocean surface in response to a breeze over the ocean. Music on a grander scale includes majestic meanders of the Gulf Stream that can appear as spontaneously as the undulations of the Jet Stream discussed in chapter 7. The meanders can become so voluptuous that they loop back on themselves, creating rings that are pinched off and drift away. Such a gentle whirlpool of cold water, and the marine life it supports, can persist in a warm environment for several months before rejoining the Gulf Stream. These are some of the phenomena oceanographers try to reproduce by means of their models. An invaluable test for these models is simulation of the similarities and differences between the three tropical oceans when they adjust to different changes in the winds.

The seasonal reversal of the monsoons over the Indian Ocean induces a seasonal reversal of the intense Somali current off the East African coast. Sailors have been aware of this since the Middle Ages at least and scheduled their voyages between Africa and India to take advantage of the favorable currents. Oceanographers take a keen interest in the Somali current because it is similar to other intense currents off the eastern coasts of continents. It sheds light on the dynamics of the Gulf Stream and Kuroshio, for example. In the adjustment of such currents to a change in the winds, planetary waves that propagate across the ocean basin play a central role. In any bucket of water that is disturbed, waves readily start to slosh back and forth, making it difficult to carry a full bucket without spilling any water. Fluctuations in the intensity of the winds affect the oceans similarly by exciting waves. The ocean is a thin, spherical shell of fluid on a rotating sphere and, in addition to the waves seen on the ocean surface when the wind is blowing, there are other waves that have a large amplitude, not at the surface, but along the subsurface thermocline. Some are confined to the neighborhood of the equator where they travel exceptionally fast. Because of those waves, warm surface waters can be redistributed along the equator far more rapidly than is possible at higher latitudes, which is why large-scale changes in sea surface temperature patterns, associated with a horizontal redistribution of upper ocean water, occur more readily in the tropics than in higher latitudes. Such an occurrence that leads to high sea surface temperatures in the eastern tropical Pacific is known as El Niño. The interval between El Niño episodes, three years approximately, depends on the time it takes the waves to propagate across the basin and hence depends on the width of the Pacific. (We return to this matter in chapter 9.)

A study of El Niño, and of oceanic variability in general, requires repeated measurements in the same region. Instruments in fixed locations and tide-gauges on islands, provide such information, as do measurements from commercial vessels that repeat their tracks at regular intervals. Oceanographers have equipped many of those vessels with instruments that are easy to operate, and crew members obligingly obtain valuable time-series of subsurface temperatures. Commercial ship tracks, unfortunately, are almost as few and far between as islands, especially in the southern ocean. Oceanographers have therefore developed instruments that can be left unattended for extended periods. Some drift with the surface currents and are located daily by satellite. Others are tethered to cables between anchors on the ocean floor and floats on the ocean surface. Because these moored instruments are expensive, and because the ocean is vast, they are used in clusters as part of programs to study specific phenomena in certain regions. To complement these measurements, oceanographers in the 1960s started to coordinate multiship, international programs to study phenomena such as Gulf Stream meanders, coastal upwelling off Oregon, Peru, and northwestern Africa, and the response of the tropical oceans to the fluctuating winds that drive them. The measurements motivated theories and models to explain and simulate the observed phenomena. The most realistic models, known as General Circulation Models, require supercomputers and are analogous to the atmospheric models described in chapter 7. Given an accurate description of the winds over the ocean, the models are capable of realistically reproducing phenomena such as meanders of the Gulf Stream and the fluctuations between El Niño and La Niña. In 1982 it became clear that this expertise needed to be put to practical use, that it was time for a new era of oceanography.

When a group of experts met in Princeton, New Jersey in September 1982, to plan a large international program to study El Niño, none was aware that the most intense episode of the century was underway and none anticipated that, during the subsequent months, it would be associated with a series of natural disasters, on a global scale. Those disasters—the floods that devastated Ecuador and Peru; the numerous storms that battered the coast of California and those of the Gulf states; the exceptional hurricanes that threatened and damaged usually tranquil islands of the tropical Pacific such as Tahiti; the droughts and forest fires of New Guinea, northern Australia, and southern Africa—received extensive coverage in the press and on television, underlining the need to alert the public of these hazards in a timely manner. This led to an expansion of efforts to monitor the oceans. Scientists deployed unattended instruments that relay data by satellite to certain centers where sophisticated computer models of the ocean, the counterparts of those used for weather prediction, assimilate and interpolate the data. Sensors on satellites now provide global surveys of sea surface temperatures and of variations in the height of the sea surface. Each month, information about the continually changing oceanic conditions is available in the form of maps, the equivalent of daily weather maps. The likelihood of a major El Niño developing unnoticed is now negligible.

Earth's climate depends on interactions between the ocean, the atmosphere, the biosphere, and various land and ice surfaces. To anticipate climate fluctuations and climate changes, oceanographers have to join a community of scientists whose interests cover a broad range of topics. This communal effort promises to be exciting and rewarding. It has just begun.

9

EL NIÑO, LA NIÑA, AND THE

SOUTHERN OSCILLATION

EL NIÑO affects everyone, either directly because of its influence on climate and weather, primarily in the regions indicated in figure 9.1, or indirectly because of its influence on the global economy. The extensive coverage that the press devoted to the exceptionally intense El Niño of 1982–83 impressed on everyone how diverse the impact of this phenomenon can be: devastating floods in Ecuador and Peru, where a warming of the surface waters of the eastern tropical Pacific—the signature of El Niño—is associated with the disappearance of the usually abundant fish; disastrous droughts in the "maritime" continent of southeastern Asia and northern Australia; unusual weather patterns over North and South America; poor monsoons over India; and low rainfall over southeastern Africa. These changes in climatic conditions can significantly reduce the harvests of coconuts in the Philippines and of anchovies off Peru, thus causing increases in the prices of soaps and detergents with coconut oil as an ingredient, of fishmeal fed to chickens, and of soybeans that can provide a substitute for fishmeal. The importance of El Niño is such that magazines ranging from *Reader's Digest* to *Soybean Digest* keep their readers informed about this phenomenon.

Today we anticipate El Niño with dread. There was, however, a time when it was welcomed as a blessing, as is evident from its name, which is Spanish for Child Jesus. At first this name was given to the warm, seasonal current that appears off the coast of Peru around Christmas when it moderates the low temperatures of the eastern tropical Pacific Ocean. Every few years this current is more intense than normal, penetrates unusually far south, and brings heavy rains to the otherwise arid coastal zones of Peru and Ecuador. Such an occurrence, now known as El Niño, was originally known as an *año de abundancia*, or year of abundance, when, according to an early observer, "the sea is full of wonders, the land even more so. First of all the desert becomes a garden. . . . The soil is soaked by the heavy downpour, and within a few weeks the whole country is covered by

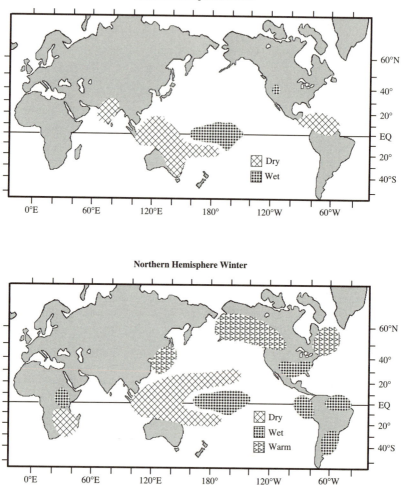

Figure 9.1 Regions affected directly by El Niño.

abundant pasture. The natural increase of flocks is practically doubled and cotton can be grown in places where in other years vegetation seems impossible." The wonders in the sea can sometimes include long yellow-and-black water snakes, bananas, and coconuts carried southward by the warm current from the coastal rain forests farther north. At the same time, however, the birds and marine life that are usually abundant temporarily disappear.

Today heavy rains can still transform parts of the Peruvian desert into a garden, but they also wash away large numbers of houses and roads. The sea is still full of wonders, but those exotica no longer compensate for the disappearance of the fish and birds (guano providers) on which the economies of Ecuador and Peru have become dependent. Because rapid population growth has increased our vulnerability to climate fluctuations, we now view El Niño pejoratively. (The term no longer refers to the seasonal coastal current, but is now reserved for the interannual *años de abundancia.*)

Not only the public, but scientists also have changed their perspective of El Niño during the twentieth century. For a long time the appearance of unusually warm surface waters off the coasts of Peru and Ecuador was thought to be a local, coastal phenomenon. Only in the 1950s did scientists acquire data that revealed a different picture. By that time, traffic across the Pacific was so heavy that measurements from commercial ships started to provide information about changes in large-scale sea surface temperature patterns. Furthermore, scientists organized for 1957 an International Geophysical Year of intensive measurements of our planet. It so happens that El Niño occurred in 1957. The data gathered during that year revealed that the unusually warm surface water was not confined to the coasts of Ecuador and Peru but extended thousands of kilometers offshore and covered much of the eastern tropical Pacific. The data also showed that the trade winds over the Pacific were unusually relaxed that year. Jacob Bjerknes, an atmospheric scientist at the University of California, Los Angeles proposed that the change in the winds caused the change in sea surface temperature patterns. Why did the winds change?

Intensive studies of variations in the atmospheric circulation of the tropics, and of the winds over the Pacific, started shortly after Sir Gilbert Walker became Director General of Observatories in India in 1904. Walker arrived just after the failure of the monsoons in 1899 that contributed to a catastrophic famine. He wished to understand the monsoons better and, in the belief that poor monsoons must be part of a global climate fluctuation rather than a local Indian phenomenon, searched for global patterns in meteorological data from around the world. Through his search, Walker discovered the Southern Oscillation, a coherent, interannual fluctuation in atmospheric conditions, that amounts to a see-saw across the Pacific: when pressure is high in the Pacific Ocean, it tends to be low in the Indian Ocean from Africa to Australia. An increase in pressure at Tahiti therefore coincides with a decrease at Darwin in northern Australia, as is evident in figure 9.2. That figure also shows that a cycle, on the average, lasts three to four years. Walker documented the associated variations in the winds and rainfall patterns and found that the sporadic failure of the monsoons

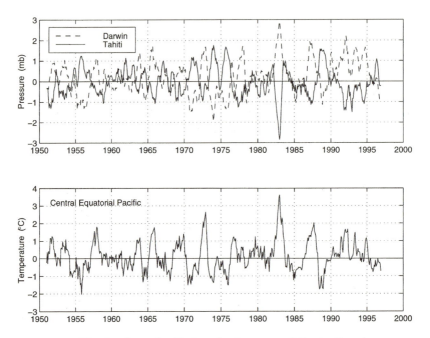

Figure 9.2 The Southern Oscillation, the interannual fluctuations in pressure that are out of phase at Darwin (Australia) and Tahiti. These variations are highly coherent with those in sea surface temperature in the eastern equatorial Pacific. (The unit for the departure of sea surface temperature from 22°C is °C.)

often coincides with low pressure over Tahiti, high pressure over Darwin, and relaxed trade winds over the Pacific. Unfortunately, Walker was unable to translate this result into a scheme that predicts failures of the monsoons. Furthermore, colleagues questioned his results; he could not explain the processes that give rise to coherent global oscillations with a timescale of several years. As a consequence, Walker's results fell into oblivion for several decades.

Interest in Walker's work revived in the 1950s when Bjerknes, on the basis of the sea surface temperature data mentioned earlier—especially data gathered during El Niño of 1957—found a connection between the Southern Oscillation and El Niño. Bjerknes proposed that the relaxed trade winds of 1957 not only caused the appearance of unusually warm surface waters over the tropical Pacific, but were in turn caused by the increase in sea surface temperatures during that year. This tantalizingly circular argument implied that the phenomenon was neither strictly atmospheric nor strictly oceanic, but was a product of interactions between the two media.

Bjerknes next generalized brilliantly and argued that the continual Southern Oscillation, not just the event of 1957, is both the cause and the consequence of continually changing sea surface temperature patterns. This meant that El Niño, rather than an isolated event that occurs sporadically, is but one phase of a cycle. In the same way that the seasonal cycle is an oscillation between winter and summer, so the Southern Oscillation is a fluctuation between El Niño and a complementary state, which has been given the apposite name La Niña. Whereas the seasonal cycle is forced by regular variations in the intensity of sunlight, the Southern Oscillation corresponds to a natural mode of oscillation of the coupled ocean and atmosphere and is literally the music of our spheres (the atmosphere and hydrosphere).

El Niño and the Southern Oscillation, which for a long time were regarded as independent oceanic and atmospheric phenomena, are different facets of the same climate fluctuation that involves interactions between the ocean and atmosphere. Bjerknes came to this realization in the 1950s; only after the unanticipated and devastating El Niño of 1982–83 did scientists quantify his ideas by developing coupled ocean-atmosphere models capable of predicting El Niño. Since then, progress has been rapid, but several puzzles remain to be solved. The unexpected persistence of warm El Niño conditions during the early 1990s was not anticipated by any of the models that successfully predicted El Niño during the latter part of the 1980s. Inspection of records for the past century reveals that during some decades, those of the 1920s and 1930s, El Niño can almost be absent. Scientists are currently exploring the reasons for the variable intensity and irregular visits of El Niño.

The Southern Oscillation is but one example of a climate fluctuation; it stems primarily from interactions between the atmosphere and the tropical Pacific Ocean. Other examples of climate fluctuations— such as a series of exceptionally severe winters over northern America or prolonged droughts over Africa and other parts of the world— probably involve additional interactions between the atmosphere and the water, land, and ice surfaces beneath it. The tools that are being developed to predict the Southern Oscillation—climate models and a network of instruments that continually provide a description of the atmosphere, oceans, land surfaces, and ice volumes—are the same as the tools that are needed to predict climate variability in general. Anticipating El Niño is therefore the first step toward operational prediction of longer-term climate fluctuations, the complement of daily weather forecasts.

The Southern Oscillation

The tropics have three main regions with heavy rainfall that sustains lush tropical jungles: the Amazon and Congo River basins and the "maritime continent" of the western Pacific, southeastern Asia, and northern Australia. The massive, tall cumulus clouds over these vast regions are the furnaces that drive the atmospheric circulation of the tropics; the fuel is the latent heat released by condensation of water vapor in the clouds. That heat makes the air buoyant, causing it to rise. To sustain the rising motion, winds near Earth's surface converge onto these regions while the air aloft, drained of its moisture, diverges from them. In the tropical Pacific, the air aloft moves eastward, subsides over the cold waters off the western coast of the Americas, and returns westward in the trade winds, acquiring moisture by means of evaporation while doing so. This is known as the Walker circulation in the Pacific; the Atlantic has a counterpart.

The locations of the furnaces, the convective zones of rising air and low surface pressures, are determined by temperature patterns at Earth's surface. The air ascends where surface temperatures have maxima. The seasonal north-south migrations of the convective zones therefore tend to keep those zones in the summer hemisphere. Over Africa and South America the zones of heavy rainfall are difficult to dislodge from the continents because surface temperatures can attain higher values on land than over the oceans. The maritime continent of southeastern Asia is an entirely different matter because its eastern boundary coincides with that of the pool of warm water that covers the western tropical Pacific. Should this pool expand eastward, so would the region of rising air and heavy rainfall, which is exactly what happens interannually during El Niño. On such occasions, the eastern tropical Pacific experiences an increase in sea surface temperatures and in rainfall, a decrease in surface pressure, and a relaxation of the trade winds. Because of this eastward shift, the tropical regions west of the date line, including India and southeastern Africa, experience decreases in rainfall (fig. 9.3a).

Figure 9.3b shows conditions during La Niña; they are complementary to those during El Niño. The pool of warm surface waters contracts westward and, over the eastern equatorial Pacific, the trade winds intensify and rainfall decreases.

Confirmation that the Southern Oscillation is caused by the interan-

La Niña Conditions

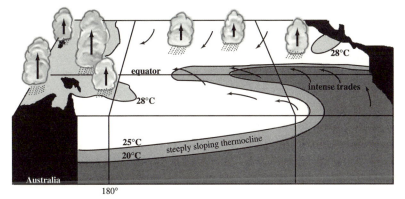

El Niño Conditions

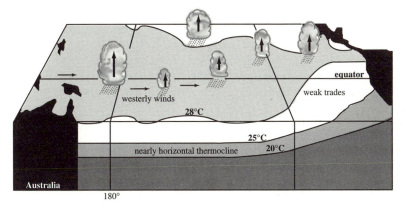

Figure 9.3 A schematic view of La Niña (*top*) and El Niño (*bottom*) conditions. During La Niña, intense trade winds cause the thermocline to have a pronounced slope from east to west so that the equatorial Pacific is cold in the east, warm in the west where moist air rises into cumulus towers. The air subsides in the east, a region with little rainfall except in the doldrums where the northeast and southeast trades converge. During El Niño, the trades along the equator relax, as does the slope of the thermocline when the warm surface waters flow eastward. The change in sea surface temperatures is associated with an eastward shift in the region of rising air and heavy precipitation.

nual changes in sea surface temperature patterns, the continual contraction and expansion of the pool of warm surface waters that covers the western tropical Pacific, comes from the computer models of the atmosphere that are used to predict weather and to simulate climate. Those models reproduce the Southern Oscillation realistically, provided the observed changes in sea surface temperatures are specified. It follows that predictions of El Niño and La Niña are possible indefinitely into the future provided we know how sea surface temperatures will change.

Oceanic Adjustment

The westward trade winds drive the warm waters of the tropical Pacific Ocean westward while exposing colder water from below to the surface in the east. During El Niño, the trades relax and the warm water surges eastward; during La Niña, the trades intensify and drive the warm water back to the west. Thus, the interannual changes in sea surface temperatures are associated with a horizontal redistribution of the warm waters of the upper ocean in response to changes in the winds. This redistribution is evident in figure 9.3, where the slope of the thermocline, the layer of large vertical temperature gradients that separates the warm surface water from the cold water below it, is seen to change significantly between El Niño and La Niña.

Tide-gauge measurements confirm that El Niño is associated with an enormous horizontal redistribution of warm waters, from the western to the eastern tropical Pacific. Recall that the height of a column of water in the ocean depends on the average temperature of the column. In the west, the height is at a maximum during periods of intense trades when the upper part of the column is a deep layer of very warm water. When some of that warm water flows eastward during El Niño, then the average temperature of the water column in the west decreases and so does the height of the column of water. The drop in sea level in the western tropical Pacific during El Niño can be so large that coral reefs become exposed and suffer damage. It has been suggested that El Niño contributed to some military disasters during World War II. The allies, when they invaded certain islands, found that their boats ran aground far from the shore, making them easy targets. Sea level was unexpectedly low because of El Niño.

The fall in sea level in the western tropical Pacific during El Niño, because of the eastward flow of warm surface waters, is accompanied by a rise in sea level in the eastern tropical Pacific. The eastward currents in figure 9.3 intensify during El Nino, whereas the westward currents weaken and sometimes even disappear. Because of the east-ward transport of unusually large amounts of warm water, the cold waters that are rich in nutrients are no longer at the surface in the east. This change in oceanic conditions is the reason for the disap-pearance of the cold-water fish along the shores of Peru and Ecuador. The cold water, and the fish, often recede southward so that Chile benefits from El Niño.

The models that oceanographers have developed to study and sim-ulate the oceanic adjustment to changes in the surface winds repro-duce interannual sea surface temperature changes realistically, pro-vided the winds are specified. These models are the counterparts of the ones atmospheric scientists use to predict weather. Together, the oceanic and atmospheric models should be capable of reproducing the Southern Oscillation by simulating interactions between the ocean and atmosphere.

Interactions between Ocean and Atmosphere

From an atmospheric point of view, the changes in rainfall patterns, winds and surface pressures associated with the Southern Oscillation are caused by changes in sea surface temperature patterns. From an oceanic point of view, those changes in sea surface temperature pat-terns are a consequence of changes in the winds. The ocean-atmo-sphere interactions implied by this circular argument are illustrated in figure 9.4 for the case of El Niño of 1982–83. Initially, unusually weak trade winds appeared over the far western tropical Pacific Ocean. Those westward winds drove the warm surface waters westward so that their weakening caused an eastward expansion of the pool of warm surface waters, which in turn caused an eastward migration of the region of rising air and heavy rainfall. An initial, cautious retreat by the trades was answered with a tentative eastward step by the warm surface waters. This response quickened the retreat, it caused a further reduction in the intensity of the trades, which emboldened the pursuit. The warm surface waters and humid air therefore surged across the tropical Pacific and soon they were hugging the shores of Latin America. El Niño had arrived.

Once El Niño is established, the stage is set for La Niña to make its entrance. This new phase of the Southern Oscillation is an inversion, a mirror image, of the first part of this duet for ocean and atmo-sphere. Consider the consequences of an initial, slight intensification of the trades that causes a slight increase in the westward flow of the warm surface waters of the tropical Pacific. Because more cold water

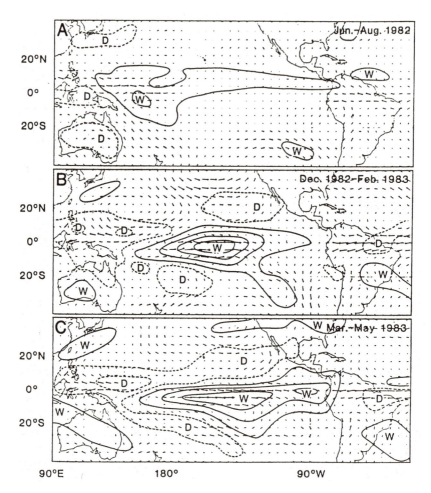

Figure 9.4 In 1982, El Niño conditions first appeared to the west of the date line and gradually expanded eastward. The panels show how, along the equator, the easterly trade winds collapsed and were replaced by westerly winds, while the zone of heavy precipitation migrated eastward. The arrows indicate the anomalous winds (the departure from the expected winds). Precipitation is unusually high where contours are solid, and is unusually low where contours are dotted. D indicates dry; W, wet. From Rasmussen and Wallace (1983).

becomes exposed to the surface in the eastern tropical Pacific, the temperature contrast between the east and west is now enhanced. That temperature difference drives the trade winds so that the initial modest strengthening of the trades now becomes magnified. This reaction goads the ocean into a stronger response, an even larger temperature contrast between the eastern and western tropical Pacific. La Niña is now in place, in due course to be followed by El Niño.

The partners in this dance are the atmosphere and ocean. But who leads? Which one initiates the eastward surge of warm water that ends La Niña and starts El Niño? Though intimately coupled, the ocean and atmosphere do not form a perfectly symmetrical pair. Whereas the atmosphere is quick and agile and responds nimbly to hints from the ocean, the ocean is ponderous and cumbersome and takes a long time to adjust to a change in the winds. The atmosphere responds to altered sea surface temperature patterns within a matter of days or weeks; the ocean has far more inertia and takes months to reach a new equilibrium. The state of the ocean at any time is not simply determined by the winds at that time because the ocean is still adjusting to and has a memory of earlier winds, a memory in the form of waves below the ocean surface. These waves propagate along the thermocline, the interface that separates warm surface waters from the deeper cold water, elevating it in some places, deepening it in others. These vertical displacements of the thermocline have little effect on sea surface temperatures in the western tropical Pacific where the thermocline is deep. Matters are different in the eastern tropical Pacific, where the arrival of a ridge or trough in the thermocline can have a profound effect on sea surface temperatures. During one phase of the Southern Oscillation, La Niña, say, the winds generate prominent troughs in the thermocline off the equator in the western Pacific; the winds in effect plant seeds that later blossom into El Niño because those troughs slowly travel across the Pacific as waves, which, when they reach the eastern equatorial Pacific, increase sea surface temperatures and signal the termination of La Niña and the onset of El Niño. During El Niño, the winds plant the seeds for La Niña in the form of thermocline ridges in the western tropical Pacific, thus ensuring a continual oscillation.

These arguments imply that the period of the Southern Oscillation, the time between one El Niño episode and the next (three to four years) depends on the time it takes certain oceanic waves to propagate across the Pacific and hence depends on the width of that basin. If so, then a similar phenomenon in the much smaller Atlantic Ocean, if there is such a phenomenon, should have a shorter period. The Atlantic does indeed have its own El Niño, which brings warm surface waters to the ordinarily cold eastern tropical Atlantic. The southwestern coast of Africa, usually a barren desert similar to the coastal zone of Peru, then has plentiful rainfall. El Niño in the Atlantic is similar to its counterpart in the Pacific except that it is far more sporadic and has a smaller amplitude. Scientists exploit these similarities and differences to test their theories.

Predicting El Niño

The continual oscillation between El Niño and La Niña described above is perfectly regular in an idealized world without any random disturbances. In figure 9.5b, which depicts such an oscillation, the warm water of the western Pacific is seen to surge eastward during El Niño and to retreat westward during La Niña. The results are from a relatively simple coupled ocean-atmosphere model that can easily be modified to explore other possibilities.

The seeds for El Niño are already present during La Niña. Suppose we modify conditions in such a way that gestation of the seeds is inhibited. This will happen should the depth of the tropical thermocline increase everywhere so that undulations of the thermocline have only a modest effect on sea surface temperatures. Under such conditions the first El Niño is followed by a weak La Niña which, in turn, is followed by an even weaker El Niño. The oscillations now decay, and finally die out altogether, as shown in figure 9.5a. Consider next altered conditions that promote gestation. Oscillations then grow to such a large amplitude that a cascade to turbulence commences, resulting in the complicated patterns of figure 9.5c.

Any natural mode of oscillation is capable of the range of possibilities in figure 9.5. Consider, for instance, meanders of the Jet Stream that depend on the temperature difference between the equator and the poles. When that difference is very small then the meanders are feeble and correspond to figure 9.5a. When the temperature difference is large, as it is in reality, then the meanders attain an amplitude even larger than those in figure 9.5c. As a result, weather patterns are chaotic and are difficult to predict.

The Southern Oscillation is not as complex a phenomenon as weather because several processes limit its amplitude. (One of those processes is the increased efficiency with which water evaporates from the ocean as temperature increases; this process prevents sea surface temperatures from rising much above 30°C.) In an idealized world without random disturbances, the Southern Oscillation would probably correspond to figure 9.5b. In such a world, El Niño, like the tides, would be perfectly predictable! In reality, depicted in figure 9.2, the Southern Oscillation is irregular because of the presence of random perturbations that include weather disturbances from outside the tropics. Those disturbances introduce an element of chaos but, as is evident in figure 9.2, the oscillation nonetheless retains much of its basic periodicity and therefore has considerable predictability. In a year during which El Niño occurs, it is possible to predict, with a

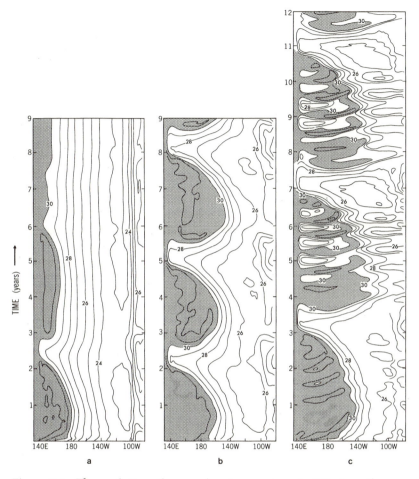

Figure 9.5 The evolution of sea surface temperature (in °C) along the equator in the coupled ocean-atmosphere model of Neelin (1990). The strength of the coupling between the ocean and atmosphere increases from (a) to (b) to (c). Regions warmer than 30°C are shaded.

modest probability of success, that La Niña will occur the next year and that El Niño will recur three years later. Statisticians can do even better on the basis of past records such as the one in figure 9.2. They have demonstrated skill in anticipating the onset of El Niño approximately a year in advance. Such methods permit predictions that El Niño will occur. The information is very useful but nonetheless is of limited value for the following reasons.

The Southern Oscillation is an irregular fluctuation with a variable amplitude that develops in a manner that changes from time to time. Sometimes El Niño starts as a modest warming of the surface waters off the coast of Peru and, over a period of many months, amplifies and propagates westward across the equatorial Pacific. That was the case in 1972, for example. On certain occasions, including 1982, it has started in the western equatorial Pacific and propagated eastward. Because each El Niño is distinct, a prediction that simply states that El Niño will occur is of limited value unless it describes how it will evolve and what amplitude it will attain. (The terms El Niño, La Niña, and Southern Oscillation are useful in the same way that the term winter is useful, even though each winter is distinct.)

Ideally we would like to have predictions that describe the expected evolution, over several months, of the rainfall patterns, winds, sea surface temperature patterns, etc. This should be possible by means of coupled ocean-atmosphere models. Given an accurate description of the initial states of the atmosphere and ocean, and assuming that the atmosphere remains unchanged for a while, the oceanic component of a coupled model can be used to forecast how the state of the ocean, especially the sea surface temperatures, will change during the next several hours. The atmospheric model can be used next to determine how the predicted change in sea surface temperatures will affect the atmosphere, especially its winds. The new winds then drive the ocean model to determine new sea surface temperature patterns, which, in turn, determine new atmospheric conditions. The models interact in this manner to produce a forecast of how the Southern Oscillation will develop. As in the case of weather forecasting, inaccuracies in initial conditions gradually degrade the forecast. Relatively simple coupled ocean-atmosphere models of this type have already been developed and can predict El Niño with a skill at least comparable to that of the statistical models that rely only on past records. Rapid progress is being made with the development of sophisticated coupled ocean-atmosphere models capable of predicting how various fields (temperature, rainfall patterns, etc.) will evolve over extended periods.

Modulation of El Niño

Interest in El Niño waned for several decades after the publication of Walker's pioneering studies in the 1930s. Not until the late 1950s did scientists again pay much attention to this phenomenon. Part of the reason is a decadal modulation of El Niño that causes irregular, prolonged periods during which it is either absent or particularly energetic; it was practically absent for a few decades early in the twentieth century.

Changes in the properties of the tropical thermocline, which separates warm surface waters from the colder water below, can contribute to long-term variations in the frequency of El Niño appearances. The thermocline is maintained by a shallow meridional circulation that links the tropical and extratropical oceans. Chapter 8 describes how water that wells up at the equator flows poleward in the surface layers, to the neighborhood of 30°N and 30°S, where the water sinks and returns equatorward in the thermocline. If a change in the tropical-extratropical exchange were to increase the depth of the thermocline, to 200 m, say, then the thermocline would be too deep for variations in its depth to affect sea surface temperatures. Such a change would induce prolonged El Niño conditions, because the sporadic occurrence of this phenomenon depends on vertical movements of the thermocline that affect sea surface temperatures.

The tropics and extratropics are so closely linked by both atmospheric and oceanic processes that large changes in climatic conditions, even if initially confined to the extratropics, will quickly (within a decade or two) alter conditions in the tropics. The cycle of ice ages—the fluctuations from glacial to warm interglacial and back to glacial conditions—almost certainly involves, not just the high latitudes that experience the biggest changes in ice volume, but the tropics, too. Such changes are bound to affect the frequency and intensity of El Niño episodes. Scientists are currently exploring such aspects of paleoclimates. The results will shed light on the way global warming, caused by an increase in the atmospheric concentration of greenhouse gases, is likely to affect El Niño.

PART THREE

10

THE PARADOX OF THE FAINT SUN

BUT WARM EARTH

THROUGHOUT its long history, Earth has always been hospitable to life; it has always maintained benign conditions in spite of perturbations such as a steady increase in the intensity of sunlight by some 30%, over the past four billion years. During that period, while the Earth maintained temperatures sufficiently moderate for liquid water to be plentiful, its one neighbor, Venus, grew so hot that its oceans evaporated, while the other neighbor, Mars, became frigid and its water froze.

Originally, at the birth of the solar system, the three terrestrial planets—Venus, Mars, and Earth—were similar in composition and differed primarily in their distances from the Sun and in their sizes. (Earth and Venus are comparable in size; Mars is considerably smaller.) The challenge is to explain why those two factors, size and distance from the Sun, caused such different developments on the different planets as the Sun steadily grew brighter. Globally averaged temperatures at the surface of a planet depend not only on the intensity of the incident sunlight but also on the greenhouse effect. Higher temperatures associated with brighter sunlight can either be reinforced or countered by a change in the greenhouse effect, because of a change in the composition of the atmosphere. What processes control that composition?

Many people regard the air around them as something apart from them; they view it the way they view water in a glass, for example, as a substance whose properties are unchanging (except for pollutants that humans inject into the atmosphere). In reality, the composition of our atmosphere is changing all the time because each of its gases continually cycles between a series of reservoirs, of which the atmosphere is but one. For example, the atmospheric concentration of water vapor is controlled by the hydrological cycle (see fig. 5.2), in which water enters the atmosphere by means of evaporation and leaves by means of rainfall. Similarly, a carbon cycle determines the amount of carbon dioxide in the atmosphere. To explain why the different planets responded differently to the intensification of sunlight, we have to explore biogeochemical cycles, especially the hydrological and carbon cycles.

The terrestrial planets acquired their atmospheres from volcanic eruptions that provided a variety of gases including the greenhouse gas water vapor. Although volcanoes continually injected more water vapor into our atmosphere, the increased water vapor did not cause a continual increase in Earth's greenhouse effect and hence in its surface temperature because the atmosphere eventually became saturated with water vapor and clouds appeared. Rainfall then removed water from the atmosphere, created an ocean, and established a hydrological cycle that determined the atmospheric concentration of water vapor.

To become saturated with water vapor, the atmospheric concentration of that gas has to reach a certain critical value. That value increases as temperatures rise (see fig. 5.3 and the accompanying discussion). Thus, in spite of the addition of water vapor, saturation can be avoided, provided temperatures increase sufficiently. Water vapor itself can provide the necessary warming because it is a greenhouse gas. This occurs on Venus, which is closer to the Sun than Earth is. Because the Sun shines more brightly on Venus, the greenhouse warming associated with a certain amount of water vapor in the atmosphere is greater. A runaway greenhouse effect is possible on Venus because the continual addition of water vapor to its atmosphere can cause such a rapid rise in temperature that saturation never occurs. Scientists believe that, early in the history of Venus, volcanic eruptions injected more and more water vapor into its atmosphere without causing saturation. The resulting increase in temperatures caused the planet to become so hot that the water vapor was able to rise to considerable elevations. Energetic ultraviolet photons then dissociated the water molecules into hydrogen and oxygen. The light hydrogen molecules escaped to space, the oxygen reacted with the rocks, and carbon dioxide from the volcanic eruptions accumulated in the atmosphere. Today Venus has no water. (Its clouds are composed of droplets of sulfuric acid.)

Over billions of years, Earth never became excessively hot—atmospheric temperatures were always sufficiently low for clouds to appear—or cold. The Earth was comfortably warm even when the Sun was faint. A plausible explanation is that our early atmosphere had a stronger greenhouse effect because of a higher atmospheric concentration of carbon dioxide. The carbon cycle that determines the concentration involves exchanges between the atmosphere, the biosphere, the oceans, and the solid earth. On relatively short time scales of sea-

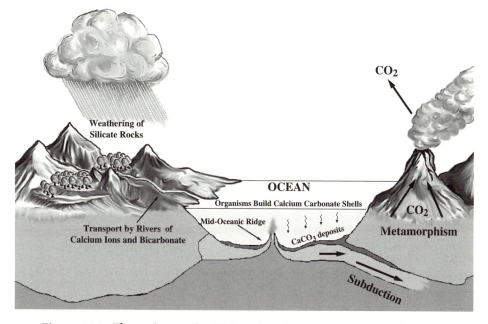

Figure 10.1 The carbon cycle. Carbon dioxide is removed from the atmosphere by means of weathering of silicate rocks. The calcium ions and bicarbonate ions thus produced are transported to the oceans by rivers. They appear in the calcium carbonate shells of certain forms of marine life that ultimately fall to the ocean floor or precipitate onto the seafloor. This sediment, carried down into Earth's hot interior in subduction zones, gives off carbon dioxide under high temperatures and pressure. That gas returns to the atmosphere in volcanic eruptions.

sons, decades, and centuries, direct atmospheric exchanges with the oceans and with the biosphere on land are dominant. For example, plants absorb carbon dioxide during the growing season and release it when they drop leaves that start to decay. Chapter 13 discusses those aspects of the carbon cycle. On the time scales of interest to us in this chapter, millions of years and longer, the most important exchange is with the sediment on the solid earth, which has a vast amount of carbon, primarily in the form of carbonate and silicate rocks, and organic matter. The carbon cycle depicted in figure 10.1 starts with carbon dioxide that enters the atmosphere by means of volcanic eruptions. The removal of that gas from the atmosphere, over millions of years, depends primarily on a process known as weathering. It involves rainwater that reacts with carbon dioxide to form a weak carbonic acid that erodes rocks containing calcium-silicate minerals. The compounds of carbon that are released from rocks are carried to the oceans by the wind and rivers and ultimately settle on the ocean floor. The presence of living organisms accelerates weathering substantially. On land, the decay of plants enhances the abundance of carbon dioxide in the soil, accelerating the conversion of certain minerals into carbonate sediments. In the oceans, some organisms use carbon compounds to build their shells, which accumulate on the ocean floor when the organisms die.

Before continuing with the rest of the carbon cycle—whereby carbon on the ocean floor returns to the atmosphere as carbon dioxide—consider the consequences of a sudden decrease in the intensity of sunlight. (This is an abrupt return to conditions during Earth's early history.) Temperatures would fall, thereby decreasing the evaporation of water into the atmosphere. The rate of rainfall, and hence the rate at which weathering removes carbon dioxide from the atmosphere, would also decrease. Lower temperatures furthermore would inhibit the biological processes that maintain high carbon dioxide levels in the soil. A fainter Sun would therefore reduce the rate of carbon dioxide removal from the atmosphere. However, volcanic eruptions would continue as before so that carbon dioxide would accumulate in the atmosphere. The enhanced greenhouse effect would then cause temperatures to rise. Thus, the drop in temperatures caused by the decrease in the intensity of sunlight would be reversed by a change in the composition of the atmosphere, by an increase in the concentration of the greenhouse gas carbon dioxide.

The negative feedback described above minimizes temperature fluctuations by means of a continual adjustment of the atmospheric composition. When temperatures are too low, the carbon cycle adjusts to increase the atmospheric concentration of carbon dioxide; when temperatures are too high, the concentration of carbon dioxide decreases because of enhanced weathering that removes carbon dioxide from the atmosphere. The carbon cycle amounts to a thermostat that keeps temperatures in a moderate range. In the past, when the Sun was faint, Earth probably managed to stay warm by accumulating more carbon dioxide in its atmosphere. As sunshine continues to grow in intensity over the next few billion years, the Earth will be able to maintain reasonable temperatures by decreasing the atmospheric concentration of carbon dioxide. In the very distant future, a billion years hence, this could create problems for the survival of plants that need carbon dioxide in the atmosphere.

The thermostat that maintains moderate temperatures on Earth failed on Venus once that planet lost its water. Once weathering ceased, carbon dioxide accumulated in the atmosphere, causing an enormous greenhouse effect. The thermostat failed on Mars, too. Inspection of the surface of that planet indicates that, at one time, it probably had liquid water; early in its history, temperatures on Mars were apparently higher than they are at present. Today Mars is so cold that its water is frozen. How is it possible that temperatures on Mars decreased over the past few billion years while the intensity of sunshine increased? Why could Earth, but not Mars, maintain comfortable temperatures?

The crucial difference between Mars and Earth is that, on Earth, the carbon that collects as sediment on the ocean floor can return to the atmosphere; on Mars, such recycling is impossible. To explain this difference we have to turn to one of the triumphs of twentieth century science, one that has dramatically altered the way we view our planet. The discovery of continental drift weaves into a unified picture seemingly unrelated phenomena such as the locations of volcanoes and earthquakes, the shape of coastlines, and the topography of the ocean floor. It furthermore enables us to anticipate more accurately where mineral and fuel resources are likely to be found, and to understand the biogeochemical cycles that are essential for the maintenance of benign conditions on Earth.

The motion of the continents depends on very high temperatures in Earth's interior, deep below its surface. Originally, the accretion of meteorites that formed Earth some 4.6 billion years ago generated a huge amount of heat that melted everything. Once in liquid form, materials segregated: very dense iron metal migrated to the center of the planet and became its solid core, which, at present, is surrounded by a liquid shell, primarily of iron but containing some nickel. This shell is covered by a mantle composed mainly of less dense silicates, magnesium, iron, calcium, aluminum, silicon, and oxygen. The outermost part of Earth is a very thin crust. This information about the interior of our planet has become available from seismology, the study of earthquakes and the motion they generate. Particularly informative are the waves that propagate away from the site (or focus) of the earthquake. They travel along paths that depend on the properties (composition, density, temperature, pressure, etc.) of the medium (rocks or liquid) through which they travel. To seismologists, earthquakes provide valuable data about Earth's interior. Why are there earthquakes? Why are they more common in some regions than others? (See fig. 10.2.)

Temperatures below Earth's surface have remained high throughout its history because of heat generated by the radioactive decay of certain elements. Near Earth's core, temperatures are so high that the metals there are fluid. At relatively shallow depths below the surface,

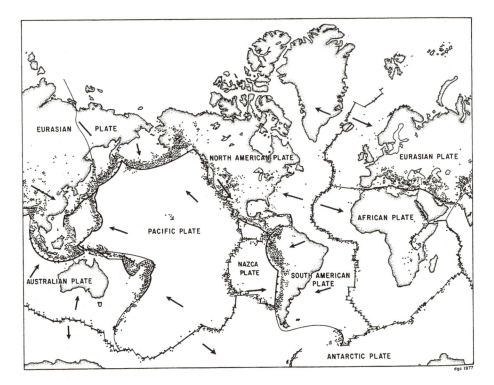

Figure 10.2 The major lithospheric plates on some of which the continents float. Arrows indicate their direction of motion. Where plates move away from each other—in the middle of the Atlantic Ocean, for example—magma wells up from Earth's interior to form midocean volcanoes, ridges, and new oceanic crust. That crust is consumed at converging plate boundaries where one plate plunges beneath another, creating volcanoes and folded mountains—along the western coast of South America, for example. Earthquakes are common along plate boundaries as is evident from the dots that indicate earthquake epicenters during the period 1961 to 1967.

temperatures are still sufficiently high for the rocks to be plastic. They deform and flow, very slowly, like an extremely viscous fluid. This motion of the plastic rocks is convective; it is driven by the high temperatures below the rocks. Earth's crust, baked from above by the Sun's heat and from below by Earth's heat, has cracked into huge plates; Earth resembles an egg with a cracked shell composed of about a dozen major plates and several minor ones (see fig. 10.2). Some of Earth's plates are entirely below the ocean, some are only partially submerged and carry the continents (fig. 10.2). These plates participate in the gradual convective motion of Earth's interior. Corroborating evidence comes from inspection of a map of the globe. A salient feature of the global map is the match in the shapes of the western coast of Africa and the eastern coast of the Americas. There is a match not only in the shapes of the coastlines of these continents but also in their rocks and in the fossils found in the rocks. On the basis of these facts, the German meteorologist Alfred Wegener proposed in 1915 that the continents were once united in a supercontinent that he called Pangaea. This idea was, for a while, rejected because it was inconsistent with certain beliefs concerning the structure of the interior of the Earth. However, accurate measurements of the depth and age of the seafloor, by means of technologies developed during World War II and in the 1950s, provided convincing evidence that Wegener was right.

The prominent features on a map of the seafloor are ridges (or mountains below the ocean surface) that stretch for thousands of miles across much of the globe. In the Atlantic, a ridge lies in the middle of the ocean basin and runs parallel to the continental margins. Surprisingly, the rocks on the ocean floor are youngest near the ridges, oldest near the continental margins. Nowhere on the sea floor are the rocks more than a few hundred million years old. New ocean floor is continually being created along the ridges where lava wells up from Earth's hot interior. The ridges, therefore, have relatively high temperatures, especially at submarine vents and in places such as Iceland which have frequent volcanic eruptions. (The vents spew hot water into the ocean and nourish bizarre creatures recently discovered by oceanographers.) The Atlantic, which acquires new crust for its seafloor from eruptions along its mid-Atlantic ridge, is expanding at a rate of a few centimeters per year, causing Africa and South America to drift apart. Some 180 million years ago, the Atlantic Ocean did not exist because the continents were merged in the supercontinent Pangaea.

If plates are drifting apart along oceanic ridges, they must be colliding elsewhere. Two plates that crash can create wrinkled mountain belts such as the Himalayas. They can also sideswipe each other, as happens along California's San Andreas fault, where accumulated stress is released in sporadic slippages or earthquakes. Another possibility is for one plate to sink beneath the other, creating oceanic trenches such as the ones off the western coast of South America. The sinking plate heats as it descends, and its material is metamorphosed; some of this material returns to Earth's surface in volcanic eruptions. The South American plate is currently sliding over the sinking plate of the eastern Pacific, causing a contraction of the Pacific and an expansion of the Atlantic Ocean.

Plate tectonics, the movements of the plates that constitute Earth's crust, effect a recycling of Earth's carbon as sediments and rocks on a moving plate sink with the plate when it is subducted. When thrust downward into the earth, the rocks and sediments are heated and transformed. Any carbon they may contain is set free, usually in the form of the gas carbon dioxide. This gas escapes to Earth's surface in the eruptions of volcanoes, or in springs that produce effervescent water that can be bottled and marketed as Perrier. Thus, carbon dioxide is restored to the atmosphere. This recycling of carbon occurs on Earth but not on Mars, because Mars's small size prevents it from having enough internal heat for plate tectonics.

The heat generated in the interior of a planet by means of radioactive decay increases its internal temperature until there is a balance between the rate at which heat is produced and the rate at which it is lost through the surface. To maximize its internal temperatures, a planet must maximize its volume, and hence its amount of radioactive material, while keeping its surface area to a minimum. Spheres are good shapes for this purpose, and large spheres are better than small ones. A small spherical planet, such as Mars, has a relatively large surface area for its volume; as a consequence, the loss of internal heat is so rapid that internal temperatures remain low. The interior of Mars is too cold for convection to be possible. Mars fails to support plate tectonics, active volcanoes, and recycling of carbon dioxide. In the case of a large planet such as Earth, internal temperatures rise to a higher value before the loss of heat through the relatively small surface balances the production. Earth can recycle its carbon dioxide and remain habitable for billions of years because it happens to be the right size.

Because it is sufficiently massive, Earth has an internal source of heat. Over billions of years, that energy not only contributed to the habitability of our planet, but it also contributed immeasurably to our civilizations by making available resources such as fuels, minerals, and metals. The average abundance on Earth of certain metals, especially the precious ones, is so low—about ten parts in a billion in the case of gold—that those metals would essentially be inaccessible if they were distributed uniformly throughout the globe. It is therefore amazing that they can be found in nuggets close to the surface. The explanation for this apparent serendipity, and also for the existence of vast, accessible reservoirs of oil, involves the high temperatures in Earth's interior.

Cold water that seeps into cracks in the rocks at Earth's surface gains heat from Earth's hot interior as it descends. The very warm water dissolves traces of metals in the rocks and carries those traces in solution over large distances. The warmer the water, the more buoyant it becomes, so that it ultimately rises. (Sometimes the hot water gushes upward through Earth's surface in geysers such as the ones in Yellowstone Park.) During this subterranean journey, changes in temperature and in the availability of oxygen (for oxidation) cause the dissolved metals to reprecipitate into veins that are easy to mine. The subsurface flow of water also assists in the creation of reservoirs of oil that has its origin in organic residue dispersed among mineral grains in sediments. It is rare for sediment to have more than 1% organic matter. However, when sediments are buried and their temperatures are increased by heat from Earth's interior, some of the organic matter converts into gas and oil that may flow through porous rocks, driven in some instances by circulating water. If these fuels collect in natural traps, they become accessible to oil drillers.

In summary, because our planet is sufficiently large, it attains sufficiently high temperatures in its interior to set the solid part of Earth in motion. This has contributed enormously to the habitability of our planet, making possible a carbon cycle that engages in a negative feedback to keep atmospheric temperatures within a moderate range while the Sun grows brighter as it grows older. This feedback is only effective if sunshine is not too intense. It fails on Venus, for example, because that planet is too close to the Sun. In due course, as the Sun shines more brightly, the feedback will fail on Earth, too, in approximately a billion years. This means that we are very fortunate. After the creation of this planet, more than four billion years elapsed before we appeared. If human evolution had taken 10% or 20% longer, we would have found ourselves on a planet about to experience a "runaway" greenhouse effect.

11

WHY SUMMER IS WARMER THAN WINTER

The Cycles of Seasons and of Ice Ages

OUR PLANET orbits the Sun with great regularity, once a year. Although its rhythm is as rigid as that of a metronome, Earth's flexible music, the weather, tends to mask the steady beat. Exhilarating accelerandos can bring summer to a sudden and early end but sometimes lazy ritardandos prolong that season, to such an extent that rhododendrons and other flowering trees bloom in October, as if summer were about to start anew. Humans cannot afford to be fooled in this manner. To make timely preparations for the inexorable march of the seasons, we have to be aware of the music's steady beat at all times, which is why we have calendars.

The misnamed last four months of the year—September, October, November, and December literally refer to the seventh, eighth, ninth, and tenth—hint at the difficulties we have encountered in trying to develop an accurate calendar. A major problem had to do with the adoption of the easily observable and reliable phases of the Moon as the basis of a calendar. The Moon completes a cycle approximately every month or 29.5 days. Twelve cycles, a lunar year, therefore last a bit more than 354 days, 11 days less than a solar year, the 365 days, 5 hours, 48 minutes, and 46 seconds it takes Earth to orbit the Sun. The discrepancy between solar and lunar years causes a purely lunar calendar, such as the Muslim one, to be inconvenient for keeping track of the seasons. The seasons drift through the year, and a specific month does not identify a specific season. Most calendars attempt to reconcile lunar and solar years. They cope with the awkwardness of months and years that involve fractions of a day by introducing intercalary days. For example, in one of the first calendars, developed by the Egyptians, a year had 12 months of 30 days each, plus five extra intercalary days, plus an extra one every four years. The Romans later adopted this calendar, but they failed to retain some of its refinements. At first the Roman year had only 10 months, the last appropriately called December, plus occasional intercalary days or months to complete the year. Later, they added two more months, January and February, and left it to the Pontifex Maximus to regulate the calendar. This power was abused for political ends, to shorten or lengthen an official's term, to such an extent that the seasons started to drift through the calendar. By the time of Julius Ceasar, autumn occurred in January. He improved matters enormously by revising the calendar, but he overestimated the length of a year slightly so that, by the sixteenth century, the vernal equinox had moved from March 21 to March 11. Pope Gregory XIII corrected this error by suppressing 10 days in the year 1582 and by ordaining that years ending in hundreds are not leap years unless they are divisible by four hundred. (The year 2000 will be a leap year.) England and the British colonies in America did not accept this reform until 1752. That is why George Washington's birthday can be given as either February 11, 1731 (Old Style) or February 22, 1732 (New Style). Today most of us use the Gregorian calendar to keep track of the passage of time and to anticipate when the seasons will change.

Accurate calendars alert us to the global climate changes that occur annually as part of the seasonal cycle. To anticipate the global climate changes that our industrial activities are likely to cause, we must rely on mathematical models of Earth's climate. We depend on models, because measurements may not reveal that climate changes are imminent. The measurements could be dominated by our planet's spontaneous music that masks impending climate changes. How do we know whether the models are reliable? Is there any information, other than that from models, about the sensitivity of Earth's climate to perturbations? To answer these questions it helps to know the reasons for the seasons.

To have seasons, a planet must meet one of two conditions: either its orbit around the sun should be an ellipse, rather than a circle, so that its distance from the sun varies during the course of a year, or the axis about which it spins should tilt relative to the plane of the orbit. In an elliptical orbit the planet enjoys summer when it is closest to the sun, at *perihelion*, and endures winter when it is farthest from the sun, at *aphelion*, provided the axis of rotation is perpendicular to the plane of the orbit. Matters are more complicated should the axis tilt. To appreciate this, consider Uranus, a planet whose axis has a huge tilt, 98°. In figure 11.1, it is evident that, when Uranus is closest to the sun, only one hemisphere gets sunshine and enjoys summer. At that time, the other hemisphere is in total darkness and has winter. When the planet is farthest from the sun, conditions on the two hemispheres are reversed so that one again has summer while the other has winter.

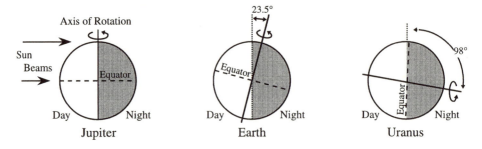

Figure 11.1 Different inclination of the axes of rotation of Jupiter, Earth, and Uranus.

Whereas Uranus is practically prostrate, Jupiter is soldierly erect because the tilt of its axis is a mere 3.1°. On Jupiter, the heat from a beam of sunshine is always most intense at the equator, where the beam is concentrated on a relatively small area. The heat is always least intense near the poles, where the beam is spread over a huge area. As the planet orbits the Sun, its distance from the Sun varies, but the equator always gets the most heat; the poles get the least. Matters are different on Earth because its axis tilts at a jaunty angle of 23.5° as it orbits the Sun. The tilt causes first one hemisphere then the other to be favored with more sunshine (fig. 11.2). During the course of a year, the latitude at which a beam of sunlight is most concentrated—the latitude at which the sun at noon is overhead—oscillates between 23.5°N and 23.5°S, latitudes determined by the tilt of Earth's axis and known as the tropics of Capricorn and Cancer. The Sun is farthest north on June 21, the summer solstice of the northern hemisphere. On that day, the northern polar cap (the region poleward of 66.5°N, a latitude determined by the tilt of the axis) experiences midnight sun, but over the southern polar cap the Sun does not rise. On December 21, summer solstice of the southern hemisphere, the southern polar cap enjoys midnight sun. At these extremes of the seasonal cycle, the solstices, Earth's axis is oriented toward the Sun so that one of the hemispheres is favored with maximum sunshine. At the equinoxes, which fall between the solstices, the orientation of Earth's axis is such that, at noon, the Sun is over the equator, and the two hemispheres receive the same amount of sunshine (see fig 11.2).

The tilt (or obliquity) of a planet's axis has a strong influence on its climate. The greater the tilt, the more sunshine the poles receive in summer. On the planet Uranus, the tilt is so large that the poles, in the course of a year, get more sunshine than the equator. If there were a planet with a tilt of 54°, then, in the course of a year, all its latitudes would receive the same amount of heat. The tilt of Earth's axis is such

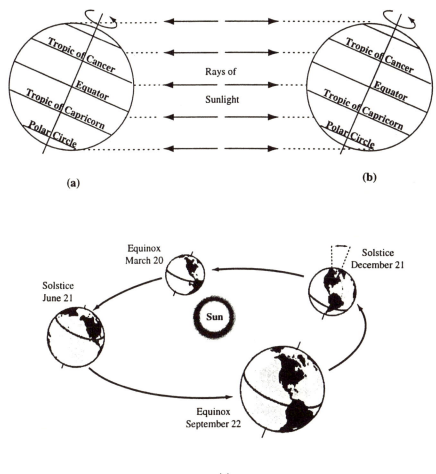

Figure 11.2 Earth's orientation on the solstice of (a) June 21, when the Sun at noon is overhead at the Tropic of Cancer and (b) December 21, when there is total darkness above 66.5°N and 24 hours of sunlight below 66.5°S. (c) Positions on Earth's orbit of the solstices, when the axis is tipped in the direction of the Sun. Aphelion, when Earth is farthest from the Sun, is close to the summer solstice of the northern hemisphere, perihelion close to the summer solstice of the southern hemisphere.

that, at present, the poles are sufficiently cold to be permanently covered with ice. If the tilt were to increase, then the poles would receive more sunshine in summer so that the polar ice sheets would probably shrink. Conversely, a decrease of the tilt would favor an expansion of those ice sheets. In reality, the tilt of Earth's axis is subject to a periodic variation: over a period of 40,000 years, it oscillates between 22.1° and 24.5°. This is but one factor that causes very gradual variations in the distribution of sunshine on Earth and hence contributes to climate changes over the millennia. Another factor is a periodic change in the eccentricity of the Earth's orbit. Over a period of 100,000 years, the shape of the orbit changes gradually from a near circle, into an ellipse, and back to a circle. Earth's distance from the Sun and the intensity of sunlight that Earth receives are practically constants during the course of a year when the orbit is almost circular, but vary when the orbit is an ellipse.

A third factor that influences the distribution of sunlight on Earth is related to the shape of our planet. Earth is not a perfect sphere but has an equatorial bulge. Because of that bulge, and because of the tilt of the axis, the Sun and the Moon force our planet to wobble. Its axis precesses and traces out a circle in space, as if it were a top spinning about a tilted axis. At present, Earth's axis points at the pole star, Ursa Polaris Minor but, 11,000 years ago, it pointed at Vega, as shown in figure 11.3. This precession does not affect the angle at which the axis tilts, only the direction in which the axis is oriented. It therefore determines where on the orbit the solstices occur; it determines which point on the orbit corresponds to June 21, which point to December 21. If Earth's orbit were a circle, this would be of no climatic significance. The orbit is an ellipse, however, and, because of the precession of the axis, we are sometimes closest to the Sun in January—that is the case at present—and sometimes we are closest in July—that was the case 11,000 years ago (see fig. 11.3). In other words, the ellipticity of the orbit favors one hemisphere with a warmer summer, but the wobble of the axis changes the favored hemisphere continuously.

Earth performs the intricate dance described above because it is under the gravitational influence not only of the Sun but also of the Moon and other planets, especially massive Jupiter and Saturn. The music for the dance has several beats. Two of them are fast, once a day and once a year. The others are very slow: a beat every 100,000 years associated with changes in the eccentricity of the orbit; another every 40,000 years as the planet rocks back and forth on its axis; the wobble of the axis (fig. 11.3) teams up with the eccentricity of the orbit to cause beats every 23,000 and 19,000 years. These periodic changes in Earth's orbital parameters can be reconstructed from Newton's law of motion which governs the motion of the planets (fig. 11.4). That is how Milutin Milankovitch calculated the past variations in sunshine in an attempt to explain a very curious phenomenon in Earth's recent past, the repeated occurrence of ice ages. How do we know that ice ages occurred?

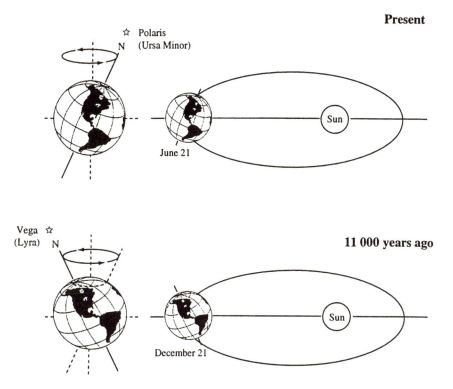

Figure 11.3 Earth's axis precesses so that the Pole Star, and also the season during which we are closest to the sun, are gradually changing. Today the Pole Star is Ursa Polaris Minor, and Earth is closest to the Sun in January. Eleven thousand years ago, the Pole Star was Vega, and we were closest to the Sun in June.

In certain parts of Europe and northern America, chunks of rocks (boulders) are found helter-skelter in locations far removed from their areas of origin. For a long time it was assumed that they had been transported in Noah's time by huge currents associated with the biblical flood. In a refinement of this theory, boulder-laden icebergs drifted about in the great flood. Further evidence that Earth had suffered a catastrophe in the past came from fossils and bones of extinct animals. Misidentification of some of the fossil bones as those of sinful people who inhabited Earth before the flood strengthened the belief that the cataclysmic event was the Deluge of Noah. How did so many of the displaced, hard, nonweathering rocks acquire deep marks and scars, and why are they often found near polished and grooved bedrock? From studies of rocks that become exposed when existing gla-

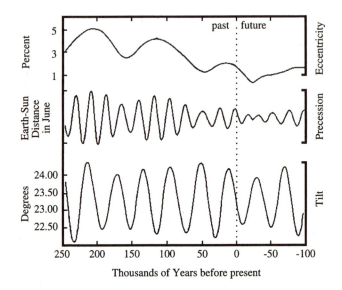

Figure 11.4 The influence of the Sun and of the other planets on Earth's orbit causes continual changes in its eccentricity, tilt, and precession that can be calculated from Newton's laws of motion. From Imbrie and Imbrie (1979).

ciers retreat, geologists concluded that the present glaciers are the modest remains of gigantic ice sheets that once covered Europe as far south as the Mediterranean and that covered large parts of North America, too. At first these results met with considerable skepticism, which geologist Louis Agassiz tried to counter with colorful prose:

> The development of these huge ice sheets must have led to the destruction of all organic life at the earth's surface. The ground of Europe, previously covered with tropical vegetation, and inhabited by herds of great elephants, enormous hippopotami, and gigantic carnivore became suddenly buried under a vast expanse of ice covering plains, lakes, seas and plateaus alike. The silence of death followed ... springs dried up, streams ceased to flow, and sunrays rising over that frozen shore ... were met only by the whistling of northern winds and the rumbling of the crevasses as they opened across the surface of that huge ocean of ice.

Glaciers expand when snowfall is so heavy that sheer weight compacts the bottom layers into ice in which boulders are imbedded. Under the force of gravity, this ice very, very slowly moves down the slopes of hills and mountains, carrying the boulders with it, scratching and polishing the bedrock over which it travels. The reverse journey is very different. The glaciers retreat when the ice melts, leaving the boulders behind, far from their place of origin.

During the nineteenth century, geologists painstakingly established the geographic extent of the glaciers during the Ice Age and determined the associated fall in sea level. So much water had been locked up in ice that sea level was more than 100 m (300 feet) lower than it is today. Such a drop exposed the seafloor between Alaska and Siberia and facilitated the trek of early Americans across the Bering Sea. In the western equatorial Pacific, it was probably possible to walk from Asia to Australia.

By the end of the nineteenth century, it was clear that Earth had experienced not one but a succession of ice ages, each separated by warmer, interglacial periods similar to the one we are enjoying now. This cycle actually can be seen on the walls of a Czechoslovakian brickyard near Nove Mesto (see fig. 11.5). The climate of that part of Central Europe, which is between the glaciers of Scandinavia to the north, and those of the Alps to the south, fluctuated as those glaciers waxed and waned, causing the boundary between the prairie and forest to march back and forth across the region. Cold, dry periods when wind-blown silt (loess) accumulated were followed by warm, wet spells when broad-leafed trees flourished and created fertile soil. The lines on the wall of the quarry mark the transitions from one climatic regime to the next.

Not only the regions between glaciers but the glaciers themselves preserve a record of past climates. By drilling into the ice sheets to obtain cores that are sometimes more than 2 km long, scientists recover snow that fell thousands of years ago in Antarctica, Greenland, and even near the equator, high in the Andes mountains. Air bubbles trapped in the ice tell us about changes in atmospheric composition. The ice itself provides information about changes in temperature. Ice consists of water molecules, each of which is composed of two hydrogen atoms and one oxygen atom. The atomic weight of oxygen is 16, but there are also a few variants with an atomic weight of 18. Some water molecules are therefore heavier than others. The lighter ones evaporate more readily from the ocean than the heavier ones. During an ice age, much of the water that evaporates from the ocean does not return as rain but becomes locked up in glaciers. Seawater therefore acquires a higher abundance of the heavy molecules, while the lighter molecules accumulate in glaciers. Scientists can establish the cycle of ice ages and the associated temperature fluctuations by measuring how the ratio of light to heavy oxygen atoms varies in the ice cores.

Cores with information about past climates can be obtained by drilling into ice sheets and also by drilling into the mud of the seafloor.

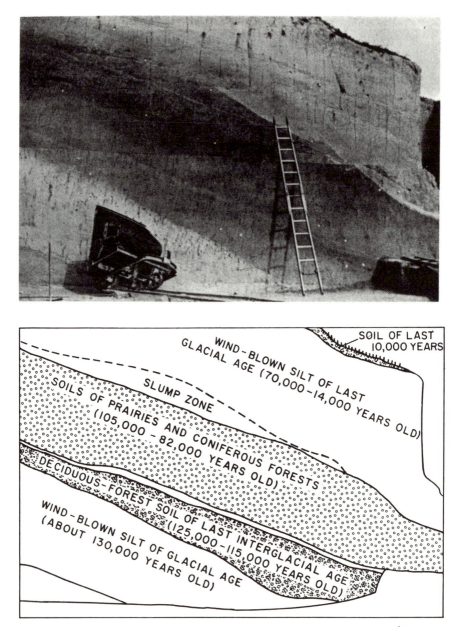

Figure 11.5 A sequence of soils and windblown silts record climatic history over the past 130,000 years in a quarry at Nove Mesto in the former Czechoslovakia. Courtesy of G. J. Kukla.

The mud is composed primarily of the remains of the inhabitants of the upper ocean, microscopic, floating organisms known as plankton. These life-forms live and die near the ocean surface where there is light. Their remains rain down on the ocean floor, covering it with an ever thickening blanket, which is thinner in the Pacific than the Atlantic because the greater acidity of the Pacific decomposes the falling remains. In temperate and tropical seas, the rain resembles snow—it is whitish—because it is rich in lime (calcium carbonate), the principal component of the skeletons of foraminifera (or forams), the most abundant planktonic animals in the warmer oceans. The polar seas have large populations of radiolaria (animals) and diatoms (plants) that extract opal, a glassy mineral, from seawater. The properties of the sediment on the ocean floor therefore change from one region to another and reflect changes in the properties of the surface waters. We can learn of different surface conditions in the past by drilling into the seafloor. The relative abundance of different species of plankton—some "cold," some "warm"—in cores taken from the ocean floor, and variations in the ratio of oxygen isotopes in those cores, tell us about past climates on Earth.

Some two million years ago, temperatures on Earth decreased significantly, to such an extent that the polar regions acquired permanent ice caps. Figure 11.6 shows fluctuations in the volume of this ice over the past 500,000 years. At present there is relatively little ice; we happen to be in a warm interglacial period. Earth last enjoyed a similar warm spell more than 120,000 years ago. Thereafter, ice started to accumulate until the volume of ice reached a peak some 18,000 years ago. From the figure, it is clear that ice accumulates very, very gradually; it melts far more rapidly. The salient feature of figure 11.6 is the recurrence of ice ages at intervals of approximately 100,000 years. It is one of the principal beats of the music of our planet. Mathematical analyses of the records reveal additional beats, at periods of 40,000, 23,000, and 19,000 years. These are exactly the periods that characterize the variations in orbital parameters shown in figure 11.4. This astonishing coincidence of results from the climate record on the one hand, and from calculations of changes in the distribution of sunshine on the other, is persuasive evidence in favor of Milankovitch's proposal that the variations in sunshine cause the recurrent ice ages. (Many puzzles do, of course remain to be solved. One concerns the dominance of a cycle of 100,000 years in the record of climate variations, but not in the record of sunshine fluctuation. Why does Earth respond most dramatically to forcing that happens to have a period of 100,000 years?)

The music of our planet, its weather and climate, has rhythm. The

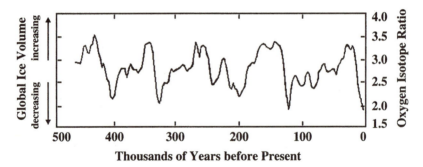

Figure 11.6 Climate record of the past 500,000 years. The measurements of the ratio of oxygen isotopes in cores from the Indian Ocean seafloor reflect variation in the volume of global ice and hence reflects global temperatures. At present the volume of ice is low; temperatures are high. Data from Hays et al. (1976).

dependable beats, those of the seasons and of the recurrent ice ages, stem from periodic variations in the distribution of sunshine. Studies of past climates therefore provide us with a wealth of information about Earth's response to precisely known variations in the factors that control its climate. The principal result is that Earth's climate is extremely sensitive to perturbations. Even though the variations in orbital parameters are very slight—the tilt, for example, fluctuates from 22.1° to 24.5°—the changes in climate are dramatic, in high latitudes where glaciers wax and wane and also in low latitudes where deserts can become verdurous during warm interglacial periods. For example, during the last warm period, some 11,000 years ago, the Middle East was verdurous and the Sahara desert had lakes. (Paleontologists have discovered the bones of hippopotami in regions that are currently too arid for those animals.) The key difference between today and 11,000 years ago is that we are closest to the Sun in January, at present, but were closest in July 11,000 years ago (see fig. 11.3). Northern summers should therefore have been warmer 11,000 years ago than they are today. The difference, though small, was sufficient to heat the African and Asian continents to such a degree that their summer monsoons intensified significantly. The winds blowing from the Atlantic Ocean onto the African continent carried so much water vapor that the region that is now the Sahara desert turned verdurous. These remarkable climate changes were in response to a slight temporal redistribution of sunshine: warmer northern summers, cooler winters.

The Sun, the Moon, and the other planets force Earth to gyrate and precess, thus causing fluctuations in the distribution of sunshine that dictate the steady beats of our planet's music. Establishing this result is a considerable scientific accomplishment, but much remains to be done. How do modest changes in the distribution of sunshine translate into enormous global climate changes? Positive feedbacks must play a role; they almost certainly involve changes in Earth's albedo and in its atmospheric concentration of greenhouse gases, the two most effective ways for altering global temperatures. Growing glaciers engage in positive feedbacks because they reflect sunlight, thus depriving Earth of heat so that temperatures fall, making possible the further growth of glaciers. Evidence for a feedback involving greenhouse gas concentrations is available from figure 11.7. The highly correlated fluctuations in temperature and atmospheric carbon dioxide levels over the past 160,000 years were determined by analyzing air bubbles trapped in glaciers over Antarctica. It is tempting to regard these results as a measure of the amount of warming to expect in response to a certain rise in carbon dioxide levels, but that is not the case because another factor, variations in the distribution of sunlight on Earth, cause both sets of fluctuations. The results tell us that Earth's initial response to the change in sunlight was reinforced by changes in carbon dioxide levels. The processes involved in this feedback, which most likely involved changes in the oceanic circulation, are currently under investigation.

In the same way that our planet's ever changing music, weather, tends to mask the steady beat of the seasons, so its longer-term climate fluctuations tend to mask the very slow beats associated with the ice ages that alternate with warm interglacial periods. This first became evident in records from cores obtained by drilling into the ice that covers much of Greenland. Samples could be dated with relative ease (and with an accuracy that far exceeds that of the methods described earlier) because the winds deposit a layer of dust on Greenland each summer; the dust is clearly visible between the snows of successive winters. The results show that numerous brief, warm periods interrupted the last Ice Age; the transition to warmer interglacial conditions was similarly interrupted by frequent frigid spells, some brief, some prolonged. The cold spell known as Younger Dryas—it was marked in Europe by a resurgence of the polar wildflower *Dryas octopetala*—occurred about 13,000 years ago and lasted for more than a thousand years. Records from locations other than Greenland indicate that, since that time, our planet's music has continued to depart irregularly from its steady beat. Significant climate variations included a warming of Europe between the years A.D. 1000 and 1400 approximately, when vineyards flourished in Britain, when plagues of

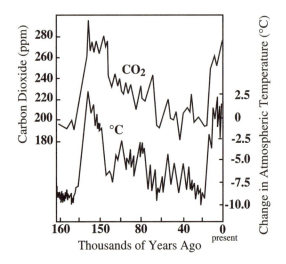

Figure 11.7 Variations in globally averaged temperatures and in the atmospheric concentration of carbon dioxide over the past 160,000 years. The carbon dioxide data come from air bubbles trapped in ice cores extracted from glaciers over Antarctica. The temperature was estimated by measuring the oxygen isotopic ratio of the frozen water. Data from Barnola et al. 1987; Lorius et al. 1990.

locusts descended on continental Europe, and when Greenland was colonized to such an extent that the Pope stationed an archbishop there. Early in the fifteenth century temperatures started to fall, forcing the abandonment of settlements in Greenland while, in Europe, cool, wet summers contributed to widespread famines. These upheavals appeared so enormous to those who experienced them that the cold spell is exaggeratedly referred to as the Little Ice Age. The very recent past has been relatively tranquil, but humans have now started to meddle. We are introducing a perturbation that is anything but modest. We are in the process of increasing the atmospheric concentration of the greenhouse gas carbon dioxide, not by a few percent, but by 100%. Will the music remain tranquil?

12

THE OZONE HOLE, A CAUTIONARY TALE

I T IS A CURIOUS STORY, about an invention that at first was viewed as a boon to mankind, then was suspected of being harmful to the atmosphere, and now is seen as a threat to life on Earth. The indictment of this invention, a family of chemicals known as chlorofluorocarbons or CFCs, came after laboratory experiments revealed that CFCs are potential vandals of our atmosphere's ozone shield. The nations of the world took swift action to prevent damage to the shield that protects us from dangerous ultraviolet rays in sunlight. In Montreal, in 1987, they agreed to limit the production of CFCs. This happened before there was any direct evidence that these extraordinarily versatile and valuable chemicals were damaging the ozone shield. When evidence of the harmful effects of CFCs did appear, it was in a dramatic and unexpected form. Whereas scientists predicted that CFCs would cause a gradual, uniform weakening of the ozone layer over many decades, a gaping hole suddenly appeared in that layer over Antarctica, where it can now be found every October. The explanation for the ozone hole involves such an unusual combination of factors that a prediction of that hole before it had been observed would have been dismissed as too improbable.

A few vocal people still refuse to accept that CFCs cause ozone depletion. They regard attempts to ban the production of CFCs as conspiracies by the chemical cartel, which has new products it wishes to market. The story of the ozone hole not only illustrates how difficult it is to predict precisely how our planet will respond when we interfere with the processes that maintain habitability, but also demonstrates how emotional the discussion of scientific matters can be.

The reason for the abundance of ozone in the stratosphere is related to the decrease in the availability of molecular oxygen with altitude—gravity keeps most oxygen molecules near Earth's surface—and the increase in the availability of atomic oxygen with altitude. Atomic oxygen (O) is created high in the atmosphere where energetic ultraviolet photons photodissociate molecular oxygen (O_2):

$$O_2 + \text{ultraviolet photon} \rightarrow O + O$$

An oxygen molecule (O_2) is converted into two oxygen atoms (O) so that a community of couples (O_2) and singles (O) evolves. Inevitably, threesomes emerge. Such an arrangement is known as ozone (O_3), an unusual molecule with three, not two atoms. The creation of ozone generates so much heat that it requires the presence of a third molecule, to be called M, to carry heat away.

$$O_2 + O + M = O_3 + M^*$$

This equation, in which M^* is a more energetic version of M, describes how ozone emerges from the collision of three particles: two molecules (O_2 and M) and an atom (O). In general, the collision of three particles is a rare event at great altitudes where the density of air is very low, but it becomes more common with decreasing height (and increasing density). Not all particles become more abundant with decreasing altitude; atomic oxygen does not because it requires the presence of energetic photons to split molecular oxygen. The elevation at which ozone can be created most efficiently is therefore a compromise between low elevations where the density of air is so high that three particles collide frequently, and great heights where one of the required particles, atomic oxygen, is abundant. This compromise is reached in the stratosphere, which therefore is the location of the ozone layer.

A threesome is not always a stable arrangement, especially not in the presence of energetic ultraviolet photons.

$$O_3 + \text{photon} \rightarrow O_2 + O$$

A photon can separate the ozone into a single oxygen atom plus an oxygen molecule. This reaction drains ozone from the stratosphere, but the reaction described by the previous equation restores ozone. The continual creation and destruction of ozone in the stratosphere, by means of the reactions described by the last two equations, maintain a stable community of singles, doubles, and triplets. The result is no net chemical change—it is as if ozone were water in a bathtub that has a running faucet to compensate for a drain that is unplugged—except for an absorption of ultraviolet radiation that leads to high temperatures in the stratosphere and safe conditions for life at Earth's surface.

This theory for the ozone layer, developed by the Englishman Sydney Chapman in 1930, explains why the highest concentration of ozone is at altitudes between 15 and 50 km, but gives values for the concentration that are larger than the observed ones. There must be additional chemical reactions that destroy ozone. In 1970, the Dutch chemist Paul Crutzen proposed that those reactions involve nitrous oxide, a chemical produced by microorganisms in the soil, and was able to explain the observed concentration of ozone in the stratosphere.

The chemicals known as CFCs, which have proven the nemesis of ozone, were invented in the 1920s. At first they were regarded as a triumph of technology because they are stable, nonreactive, long-lived, nonflammable and safe, far more so than ammonia which they replaced in refrigerators and air conditioners. CFCs turned out to have a great many additional uses: they are effective propellants in spray cans for household products and pharmaceuticals; they are good insulators and therefore standard ingredients of plastic-foam materials; and they are excellent solvents for cleaning electronic equipment. These versatile chemicals are furthermore cheap to manufacture so that production grew rapidly during the 1950s and 60s.

In the 1970s Rowland and Molina predicted that the nonreactiveness of CFCs, considered a virtue, could prove to be a major flaw. Because CFCs do not react with other chemicals, they remain in the atmosphere for a very long time, gradually dispersing upward to considerable elevations, and infiltrating the stratosphere, where they fall prey to dangerous beams in sunlight that break up the complex molecules into simpler ones. Rowland and Molina anticipated that one of the simpler chemicals, chlorine, would prove disastrous to the ozone layer. Chlorine acts as a catalyst, and a single chlorine molecule can eliminate tens of thousands of ozone molecules by participating in catalytic cycles such as the following,

$$Cl + O_3 \rightarrow ClO + O_2$$
$$ClO + O \rightarrow Cl + O_2.$$

The net result is

$$O_3 + O \rightarrow 2O_2.$$

At the end of this cycle, an ozone molecule and an oxygen atom have combined to form two oxygen molecules. A chlorine atom arranges such marriages, which remove ozone from the atmosphere, but then withdraws, remaining free to arrange more marriages, hundreds of thousands of them. The result is a thinning of the ozone shield.

Scientists projected that serious damage to the ozone layer could affect all life on Earth. It could cause a significant reduction in food production and fisheries. They also feared that an increase in ultraviolet radiation could cause a suppression of the human immune system, an increase in eye cataracts and blindness, and, most disturbingly, a significant increase in the number of deaths attributable to skin cancer.

These dire predictions persuaded the nations of the world to act swiftly. Even though, in the mid-1980s, there was no firm evidence that CFCs were causing a weakening of the ozone layer, 23 nations, under United Nations auspices, started to negotiate a protocol to limit production of CFCs. Surprisingly, the sponsors of the protocol included both environmental groups and the main chemical companies that produced CFCs in the United States. The producers had an opportunity to gauge the strength of the environmental movement in the United States when the sales of spray propellants plummeted because of the possibility that the propellants could damage the ozone layer. The companies that produce CFCs anticipated that those chemicals would be banned in the United States and realized that such unilateral action on the part of the United States would give foreign competitors an advantage. Those companies therefore favored an international agreement to ban CFCs. The first agreement, the Montreal protocol of 1987, was for a reduction in the production of CFCs but the nations wisely agreed to permit revisions should new scientific evidence make that desirable. During the negotiations that led to the protocol, the story of ozone depletion took a curious turn.

The Ozone Hole over Antarctica

Rowland and Molina predicted a gradual, global depletion of the ozone layer over the next several decades, but the first observations of ozone depletion, published in 1985, showed it to be massive and local, over Antarctica where it appears during the month of October. (The hole is absent during earlier months and disappears in November.) The measurements were so unexpected that, for a while, the cause of the ozone hole was a mystery. A field program, including flights into the stratosphere over Antarctica and laboratory experiments, established the reasons for the ozone hole. It is a consequence of chemical reactions that scientists at first ignored. Initially scientists paid attention primarily to reactions in mixtures of gases and disregarded heterogeneous reactions, those that occur on the surfaces of particles. (For two molecules to react, they have to meet. The frequency with which this happens can be low if the molecules are flying around randomly in the rarefied stratosphere, but increases should the molecules linger on the surfaces of solid particles.) Suitable particles include minute ice crystals in clouds that form high above Antarctica when temperatures are extremely low, during the long polar night. On these surfaces, certain forms of chlorine that do not react with ozone are converted into forms that do, after photo-dissociation by photons. The stage is set for a massive destruction of ozone when two requirements are met: temperatures must be so low that clouds composed of solid particles can form in the stratosphere, and photons must be available to split certain chlorine molecules. One of these requirements, low temperatures, is satisfied during the long Antarctic night, during July and August, but there are no energetic photons at that time. The other requirement, plentiful energetic photons, is satisfied during the southern summer, November and December, but, because of the relatively high temperatures once the Sun is above the horizon, the stratosphere at that time has no clouds composed of ice particles. Only in September and October, when the Sun is barely over Antarctica and when temperatures are still low, are conditions favorable for an ozone hole. These arguments imply that an ozone hole should also appear over the Arctic Ocean in March and April. At first, none was observed, but a hole did appear in March 1997.

Geographically, Antarctica is a relatively isolated continent surrounded by oceans. The winds that circle it enhance the isolation of the air over that continent and minimize the exchange of air between the high and low southern latitudes. The cold air over the less isolated North Pole, by contrast, often gets replaced by warmer air from lower latitudes and altitudes, because continents surround the Arctic Ocean and their terrain, which is far rougher than the smooth ocean around Antarctica, influences the winds in such a way as to promote an exchange of air between the polar and subpolar regions. For such reasons, an ozone hole forms over the Arctic far less readily than over the Antarctic. Over the past decade there have been occasions when temperatures in the Arctic stratosphere were sufficiently low to favor an ozone hole, but on most occasions those conditions did not persist into the period of sunlight because of an influx of warmer air into the polar vortex. In March 1997 all the conditions were satisfied, and a hole similar to that over Antarctica formed over the Arctic Ocean.

Other Explanations for Ozone Depletion

Although there is a consensus that chlorine contributes significantly to ozone depletion, there is disagreement about the principal source of chlorine in the stratosphere. Chlorine enters the atmosphere at Earth's surface in various forms: as common salt (sodium chloride) in sea spray, as hydrogen chloride in volcanic eruptions, and as a component of the complex CFC molecules. To harm the ozone layer, these chemicals must rise to the stratosphere, where energetic photons can break them into simpler components, which then attack ozone. Which forms of chlorine are most likely to rise high above Earth's surface? The salt in sea spray readily dissolves in water so that rain rapidly washes it from the troposphere, long before it can reach the stratosphere. Some people argue that CFCs, too, are unable to reach the stratosphere because they are far too dense. CFCs are indeed heavier (more dense) than air and, if introduced into a still room, will accumulate near the floor. But the atmosphere is not motionless; it has powerful fans that effectively mix light and heavy molecules until they are evenly dispersed. One of those fans is convection, which disperses parcels of air vertically. Strong updrafts in a thunderstorm can carry parcels of air, composed of light and heavy molecules, from Earth's surface to the top of the troposphere within a matter of minutes. Above the troposphere, in the stratosphere, strong winds continue to disperse the parcels. Long before CFCs became controversial, analyses of samples of air from different altitudes up to heights of 50 km and more, established that the dispersion of air molecules is independent of molecular weight. It is therefore possible for CFCs to be a source of chlorine to the stratosphere. But are they more important than volcanoes?

There is no doubt that a powerful volcanic eruption can release chlorine into the stratosphere. But how much chlorine? After the 1976 eruption of Mount Augustine in Alaska, and before there were any direct measurements in the stratosphere, attempts were made to estimate the amount of chlorine that volcanoes can provide to the stratosphere. Those estimates, based on hypotheses that seemed reasonable at the time, indicated that the 1976 volcano deposited into the stratosphere the equivalent of 17% to 36% of the industrial production of chlorine in the form of CFCs in 1975. By extrapolation, it was then estimated that the enormous volcanic eruption of 700,000 years ago, which left the Long Valley caldera in California, deposited into the stratosphere 570 times as much chlorine as was released in 1975 in the form of CFCs.

In the late 1970s scientists started measurements to determine the stratosphere's sources of chlorine. They have found that the increase in the concentration of stratospheric chlorine matches the rate of production of CFCs, and that the increase was not affected significantly by two major volcanic eruptions, El Chichon in Mexico in April 1982 and Pinatubo in the Philippines in June 1991. The measurements invalidate the hypotheses on which the earlier estimates of volcanic eruptions were based and provide convincing evidence that the principal source of chlorine for the stratosphere is CFCs. Those who argue in favor of volcanoes ignore these measurements and, on a few occasions, have committed a major error by attributing to the 1976 Alaskan volcano the massive release of chlorine estimated for the volcano of 700,000 years ago!

Although there is now convincing evidence that CFCs are damaging the ozone layer, many questions concerning the global depletion of ozone remain to be answered. For example, why, in January 1993, were ozone levels 13% to 14% below normal in northern midlatitudes? (Ozone had been thinning at an average global rate of 3% per decade.) Are scientists overlooking some important chemical reactions? The discovery of the ozone hole in the 1980s caught scientists by surprise because nobody had anticipated the important heterogeneous chemical reactions on particle surfaces mentioned earlier. Only after they learned of the ozone hole, did scientists realize that the presence of ice particles in stratospheric clouds accelerates the destruction of ozone. Sulfuric acid droplets may play a similar role; their numbers in the stratosphere increase significantly after volcanic eruptions. Could the explosion of Mount Pinatubo and its release of sulfur have caused the large ozone depletion observed in 1993? Scientists continue to explore this possibility.

Damage to the ozone layer is most severe over Antarctica. The amount of harmful ultraviolet radiation that reaches the surface of that continent has increased substantially. In other parts of the globe, where damage to the ozone layer is more modest, there are no measurements to determine whether an increase in ultraviolet radiation has reached the Earth's surface. A signal will be difficult to detect while the damage to the ozone layer is relatively modest because the intensity of ultraviolet radiation at Earth's surface has considerable natural variability. It fluctuates with the time of the day and with the seasons; the intensity is greater at noon than at sunset, and is greater in summer than in winter. (Rays from the Sun travel greater distances through the atmosphere at sunset than at noon, in winter than in summer. The greater the distance, the greater the attenuation of the rays.) The intensity of ultraviolet radiation also varies from place to place. Although they are at the same latitude, the citizens of Denver are exposed to more ultraviolet radiation than those of New York, because Denver has a greater elevation. (Rays from the Sun travel through less atmosphere to reach Denver.) Radiation is more intense in low latitudes—those of Miami, for example—than in the higher latitudes of cities such as New York. Because of all these variations in the intensity of ultraviolet radiation, it will be difficult to detect additional radiation caused by a damaged ozone layer, until the damage is severe.

The story of the ozone hole is a cautionary tale about the limited abilities of scientists to anticipate how Earth is likely to respond to perturbations. Nature's response to minuscule changes in the chemical composition of the atmosphere can be in the form of bizarre phenomena (such as the ozone hole) that are difficult, if not impossible, to predict. It is to their credit that scientists succeeded in anticipating that CFCs could harm the ozone layer, and that diplomats succeeded in negotiating a treaty limiting the production of CFCs.

13

GLOBAL WARMING, RISKY BUSINESS

EVERYBODY became familiar with the terms "global warming" and "greenhouse effect" during the horrendously hot, dry summer of 1988. Crops failed so disastrously in the midwestern United States that cattle had to be slaughtered for lack of grass to feed them. Winds swept topsoil into dark clouds on the horizon, clouds reminiscent of the Dust Bowl days of the 1930s. Newspaper articles and television reports showed pictures of barges stranded in the Mississippi River, which was running dry, and of forest fires that ravaged millions of acres in the west. In the eastern states, temperatures were so unbearably high that assembly lines were shut down in some factories. The Soviet Union and China were similarly drought-stricken, but torrential rains plagued parts of Africa, India, and Bangladesh. At one stage, three-quarters of Bangladesh was under water. In the Yucatan Peninsula, exceptionally intense Hurricane Gilbert practically swept towns into the sea. At the end of that year, the cover of *Time* magazine had a picture, not of the "Man of the Year," but of planet Earth, a planet in peril.

In televised congressional hearings, scientists sounded the alert that summers such as that of 1988 are likely to increase in frequency as a consequence of our industrial and agricultural activities, which are causing an increase in the atmospheric concentration of greenhouse gases. Figure 13.1 shows that the rate at which we are burning fossil fuels was off to a slow start, around 1850, the beginning of the Industrial Revolution, then faltered during the Great Depression of the 1930s, speeded up after World War II, and started to sprint during the 1960s. The Oil Embargo of 1973 stopped us in our tracks, but for a moment only. Since overcoming the Organization of Petroleum Exporting Countries (OPEC), in part by becoming more energy efficient, we quickly recovered momentum and are now galloping at a dizzying pace.

The summer of 1988 was followed by a few more exceptionally hot summers. Calls for governments to take action grew in urgency. In June 1992, in Rio de Janeiro, 154 nations agreed on a convention that comes into force after 50 of those nations have ratified it. One of the

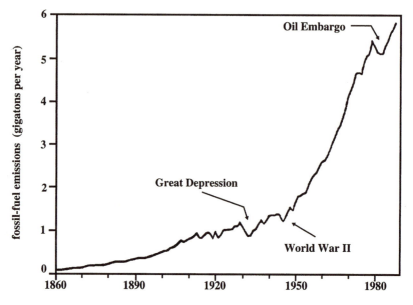

Figure 13.1   The rate at which we are injecting carbon into the atmosphere.

goals of the convention is to achieve "stabilization of greenhouse gas concentrations in the atmosphere at a level that would prevent dangerous anthropogenic interference with the climate system."

Nobody is in favor of "dangerous anthropogenic interference with the climate system." The release of greenhouse gases into the atmosphere, and pollution in general, happen to be by-products of the industrial and agricultural activities that maintain our standard of living. There's the rub. How do we weigh the possible harm of our actions against the advantages of economic growth, especially in the case of poor countries? How do we enforce regulations without limiting the freedom of individuals? In trying to cope with these difficult questions, it would be helpful if scientists could provide accurate predictions of global climate changes. A considerable effort is now underway to determine how Earth's climate will respond to increases in the atmospheric concentrations of greenhouse gases. To keep governments and the public informed of the latest scientific results, the United Nations created in 1988 an Intergovernmental Panel on Climate Change to assess the available information. This panel, whose members include hundreds of scientists from around the world, issued its first comprehensive report in 1990, another in 1996. These reports represent the views of a large number of experts, but not of everyone. The following is a brief summary of the main points of agreement and disagreement.

Points of Agreement

"Global climate changes, including global warming, are inevitable should the atmospheric concentration of greenhouse gases rise continually." The evidence in favor of this statement is both empirical and theoretical. Theories that describe the interactions between light and air (chap. 3) indicate which gases can provide a greenhouse effect by absorbing infrared radiation from Earth's surface. Satellite measurements show absorption of that radiation at exactly the wavelengths that the theory predicts. Evidence that an increase in the concentration of greenhouse gases in a planetary atmosphere is associated with an increase in surface temperatures is plentiful. The planet Venus, for example, even though it absorbs less sunlight than does Earth, has a far higher temperature because it has a much higher concentration of carbon dioxide in its atmosphere. On Earth, fluctuations in atmospheric carbon dioxide levels over the past hundred thousand years are highly correlated with variations in temperatures, as is evident in figure 11.7. Because of such evidence, scientists agree that, given a continual rise in the atmospheric concentration of greenhouse gases, global warming is inevitable. They disagree about the timing and magnitude of that warming.

The atmospheric concentration of greenhouse gases is rising. Toward the end of the nineteenth century, the Swedish chemist Arrhenius, first predicted that, because our industrial activities will increase atmospheric levels of the greenhouse gas carbon dioxide, global warming is likely. (He optimistically anticipated "ages with more equable and better climates, especially as regards the colder regions of the Earth.") Nobody paid much attention because scientists could not establish convincingly that anthropogenically produced carbon dioxide was accumulating in the atmosphere; their instruments were too crude. They furthermore assumed that the oceans, which have vastly more carbon dioxide than the atmosphere, would prevent such an accumulation by absorbing the gas. In the 1950s, Roger Revelle and Hans Suess of the Scripps Institution of Oceanography, La Jolla, California argued that there are limits to the amount of carbon dioxide that the ocean will absorb and that the atmospheric levels of that gas must be rising. Shortly afterward, their colleague Charles Keeling invented an instrument that measures atmospheric carbon dioxide levels accurately. He started to use it on Hawaii, on the slopes of a volcano in the middle of the Pacific Ocean far from industrial sources of carbon dioxide, and soon established beyond doubt that atmospheric carbon dioxide levels are increasing rapidly. The analyses of air bubbles trapped in glaciers that formed when snow fell many years ago have extended the carbon dioxide record back to the beginning of the millennium and show that the increase in atmospheric carbon dioxide levels did indeed start with the Industrial Revolution (see fig. 13.2).

Carbon dioxide is but one of the greenhouse gases whose atmospheric concentration is rising rapidly. Another, nitrous oxide, is released into the atmosphere when farmers use nitrogen-based fertilizers. Yet another greenhouse gas, methane, the marsh gas that sometimes bubbles out of wetlands and burns spontaneously as flickering blue flames, correlates with the rise in the world's population. To feed more people requires more rice paddies and more cattle, both of which are sources of methane.

Of the approximately 6 gigatons of carbon that we inject into the atmosphere each year, some 3 gigatons remain in the atmosphere. To determine what happens to the rest, we have to examine Earth's carbon cycle. We are primarily interested in what will happen over the next several decades and centuries, not the next several millennia. It is therefore possible to disregard some aspects of the carbon cycle described in chapter 10, specifically the effects of volcanic eruptions and the weathering of rocks. Those processes are important on time scales of millions of years. The exchanges that matter over the course of decades or centuries, which are depicted in figure 13.3, are primarily among the atmosphere, the oceans, and the terrestrial reservoir, the plants, and surface soils. Plants absorb carbon dioxide from the atmosphere by means of photosynthesis—especially in the growing season, the spring and early summer—and return that gas to the atmosphere by means of respiration and plant decay. (The latter is particularly important after the leaves have fallen in the autumn.) The exchange between the ocean and atmosphere depends on various factors that include the relative concentration of carbon dioxide in the two media. The colder the water of the ocean, the richer it is in carbon dioxide. When, in the equatorial Pacific, upwelling brings cold water from subsurface layers to the surface, carbon dioxide bubbles out of the ocean for the same reason that it bubbles out of champagne (or soda water); its concentration in the liquid exceeds its concentration in the surrounding air. (We are fortunate that a shallow layer of relatively warm water shields most of the deep ocean from the atmosphere. If the deep ocean were warmed up to the temperature of the surface and were allowed to exchange carbon dioxide with the air, then atmospheric carbon dioxide levels would increase by a factor of

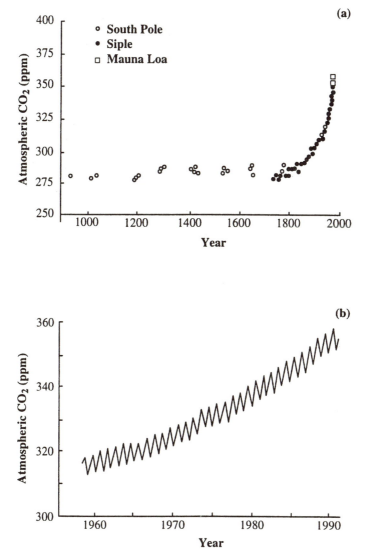

Figure 13.2 (a) Variations in the atmospheric concentration of carbon dioxide over the past 1000 years as measured in air bubbles trapped in ice cores extracted from glaciers over Antarctica. Superimposed, for the past 35 years, are direct measurements from Manua Loa, Hawaii. The latter record is shown in more detail in (b) which reveals both a seasonal cycle and a steady rise. Data from Neftel et al. (1985); Friedi et al. (1986); Siegenthaler et al. (1988); and C. D. Keeling.

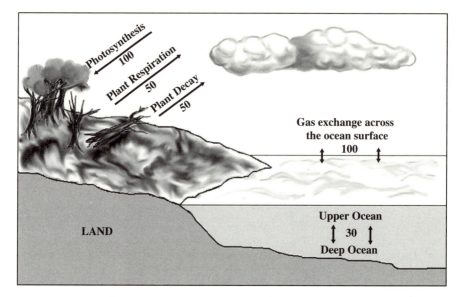

Figure 13.3 The carbon cycle involves exchanges between various reservoirs, principally the atmosphere, which has approximately 600 billion tons, the upper ocean (2,300 tons), the cold, deep ocean (35,000 tons), and the terrestrial reservoir that includes the biosphere (2,000 tons). The exchange between the atmosphere and biosphere, approximately 100 billion tons of carbon per year, is comparable to that between the atmosphere and the upper ocean, which also exchanges approximately 30 billion tons per year with the deeper, colder part of the ocean. (The amount of carbon deposited on the ocean floor in the course of a year is negligible but matters over the millennia, as discussed in chapter 10.) At present, the carbon cycle is not in equilibrium because we inject some 6 billion tons of carbon per year into the atmosphere.

2.5.) The water that wells up at the equator is heated by the Sun but then gradually cools off as it drifts poleward. The decrease in temperature increases the ability of the water to hold carbon dioxide so that it reabsorbs that gas from the atmosphere in higher latitudes. In a state of equilibrium, the net exchange across the ocean surface is zero; the ocean gains as much in high latitudes as it loses in the upwelling zones of the tropics. The net exchange between the atmosphere and the biosphere is similarly zero.

At present there is no equilibrium because we are injecting 6 gigatons of carbon (in the form of carbon dioxide) into the atmosphere each year. Some 3 gigatons remain there. The ocean absorbs approximately 1.5 gigatons of carbon per year because higher carbon dioxide levels in the atmosphere increase the flux of that gas into the ocean.

That leaves 1.5 gigatons unaccounted for. Plants on land thrive on carbon dioxide and should be absorbing more of it—tree rings could be growing fatter—but that sink of atmospheric carbon dioxide is countered by our destruction of forests, especially in the tropics. Not only are there fewer and fewer trees to absorb carbon dioxide, but by burning them, we increase atmospheric carbon dioxide levels. The mystery of the missing carbon, 1.5 gigatons per year approximately, is cause for concern. How can we be sure that the unidentified sink is not becoming saturated? If that should happen, atmospheric carbon dioxide level will soon increase more rapidly.

Determining the global climate changes in response to the greenhouse gases we inject into the atmosphere is a very complicated problem because the fraction of those gases that remains in the atmosphere changes with time; that fraction depends on biogeochemical cycles that influence and in turn are influenced by the changing climate. These complexities of the various biogeochemical cycles can be bypassed by modifying the question we ask. Rather than inquire how the climate will change in response to the injection into the atmosphere, at a specified rate, of certain greenhouse gases, we can ask how the climate will change should the concentration of greenhouse gases in the atmosphere reach a certain specified level. Scientists are therefore exploring what will happen should the atmospheric concentration of greenhouse gases increase by a certain factor—two or four, for example—over the values before the Industrial Revolution. Though simplified, the problem still remains enormously complex because the atmospheric concentration of one very important greenhouse gas, water vapor, cannot be specified. Much of the disagreement about global climate changes is about the role of water in its various guises.

Main Point of Contention

Today some people insist that the increase in global temperatures during the 1990s is clear evidence that global warming is underway. In the early 1970s, when the northern hemisphere experienced a period of falling temperatures, some of them claimed that an ice age was imminent! In discussions of the likely consequences of the steady rise in the atmospheric concentration of greenhouse gases, it is possible to make a case for any eventuality by emphasizing some feedbacks while disregarding others. The debate about global warming is primarily about the degree to which different feedbacks, many of which involve water in one form or another, will influence the Earth's climate in the future.

The term feedback describes tit-for-tats that either amplify or attenuate an initial disturbance. Those that cause amplification are said to be positive. Negative feedbacks, on the other hand, moderate the response to a perceived provocation and strive to maintain an equilibrium. An example of a positive feedback is the experience of a person who has just arrived at a cocktail party and finds that he has to raise his voice to be heard by the person next to him. This causes a nearby group to speak slightly louder in order to continue their conversation. Others in the room now have to speak even louder, and so on until nobody can hear anything. This positive feedback can also coerce everyone into silence. A group that starts to speak a bit more softly makes others feel self-conscious about being loud and initiates a feedback that, after a while, results in momentary silence in a crowded room.

In the 1970s, some people invoked certain positive feedbacks to bolster their claim that an ice age is imminent. They proposed that higher carbon dioxide levels in the atmosphere lead to higher temperatures so that more water evaporates from the oceans into the atmosphere. The winds carry the moisture poleward, where it precipitates as snow. The resulting increase in Earth's albedo causes the reflection of more sunshine so that temperatures fall. Precipitation therefore continues in the form of snow rather than rain, causing glaciers to expand even further, to reflect even more sunlight, and so on. That is how positive feedbacks can lead to an ice age!

The flaw that invalidates this argument is the isolation of one feedback from a host of competing feedbacks. By isolating a different positive feedback, one that again involves water, it is possible to argue the very opposite, that a modest increase in atmospheric carbon dioxide levels will lead, not to an ice age, but to disastrous global warming. Rather than focus on water in the form of white snow that reflects sunlight, let us consider the invisible gas water vapor that efficiently absorbs infrared radiation, thus contributing to greenhouse warming. If an initial increase in atmospheric carbon dioxide levels causes higher temperatures and therefore more evaporation from the oceans and more water vapor in the atmosphere, then the increase in temperatures is amplified. This causes even more water vapor to enter the atmosphere, resulting in further increases in temperature, and so on. Here we have the makings of a runaway greenhouse effect: it causes the atmosphere to grow continually hotter.

A runaway greenhouse effect with continually increasing temperatures is possible provided the atmosphere never becomes saturated with water vapor. If it should become saturated, then clouds will form. This brings us to yet another feedback, a negative one, in which water is involved. Water droplets in the form of a white cloud, unlike invisible water vapor, reflect sunlight extremely well. By doing so, they deprive the planet of heat. This means that clouds contribute to both global warming and cooling; their water vapor causes only heating, but their droplets cause heating and cooling. Determining the net result is no trivial matter because the size of water droplets, their concentration in a cloud, and the type of cloud (whether it is stratus or cumulus), are all factors that come into play. At some stage, negative feedbacks become dominant and arrest the increase in temperatures. Extremely important is the negative feedback that depends on the increase in the rate at which a body radiates heat as its temperature rises. (See the Stefan-Boltzman law of Appendix 3.) Another negative feedback involves clouds that reflect sunlight.

Models of the Earth's climate have to cope with numerous feedbacks simultaneously. Some critics identify specific feedbacks that are absent from the models, show that those feedbacks, in isolation, can produce results different from those in the models, and conclude that the models are worthless. This is sophistical reasoning because any feedback must be evaluated, not in isolation, but in the context of all the others, and only the models are capable of doing that. Any newly identified feedback needs to be incorporated into the models in order to determine whether it alters the results significantly. If a model that predicts global warming is found to have a flaw, it is logical to question the prediction, but it is illogical to conclude that there is no threat of global warming. It is possible that, after the flaw in the model has been corrected, the results stay essentially the same. For an objective assessment of the relative importance of the various feedbacks, we have to rely on computer models of Earth's climate (described in chap. 7) because only they can cope with such a complex problem. How do we know whether a model weighs the various feedbacks appropriately?

The best test for a model is its ability to simulate Earth's current and past climates. Although the models are remarkably successful in reproducing many aspects of those climates realistically, and are improving rapidly, critics can still readily identify unrealistic features of the simulations. Recently, the critics have been skeptical of claims that the models successfully reproduce the warming observed over the past century.

Is Global Warming Evident Yet?

The increase in atmospheric levels of greenhouse gases started 150 years ago, with the Industrial Revolution, so that it is reasonable to inspect the record of temperature fluctuations over the past century for indications of warming. The record, shown in figure 13.4, indicates that, in contrast to the steady rise in the atmospheric concentration of carbon dioxide shown in figure 13.2, temperatures have fluctuated erratically. Between the turn of the century and the 1940s, temperatures increased, but the trend then disappeared for a few decades before it reappeared in the 1980s.

Climate models, in which the concentrations of atmospheric greenhouse gases increased in accordance with the actual increases over the past century, at first reproduced global warming larger than that observed (and shown in fig. 13.4). Recently the results from the models improved significantly when scientists took into account the release into the atmosphere not only of greenhouse gases but also of aerosols, solid particles (pollutants) that reflect sunlight and thus reduce the warming associated with the greenhouse gases. The degree to which the aerosols can accumulate in the atmosphere is limited because they remain in the atmosphere for a far shorter time than do the greenhouse gases. It is therefore likely that the cooling effect of the aerosols will soon cease to increase, whereas the greenhouse warming is likely to continue increasing, thus becoming more prominent. For these, and other reasons, the United Nations International Panel on Climate Change in 1995 issued a cautious statement that "the balance of evidence suggests a discernible human influence on climate." Some scientists are skeptical.

The interpretation of a short trend in the temperature record as an indication of an imminent climate change is problematic because, in the short run, our planet's natural variability can mask trends associated with a climate change. In the same way that the onset of summer can be erratic, in spite of a steady increase in the intensity of sunshine between December and June (in the northern hemisphere), so the onset of global warming can be erratic in spite of the steady increase in atmospheric carbon dioxide levels. To determine whether the high temperatures during the 1980s and 1990s evident in figure 13.4 can be attributed to higher carbon dioxide levels, we need to (1) predict by means of a model the expected climate change signal, (2) establish the level of "noise" associated with the natural variability, and (3) calculate whether the ratio of the signal to the noise exceeds a certain threshold. Although recent improvements in climate models give scientists confidence that they can predict the expected signal, data that describe natural variability are inadequate.

Information about natural climate variability is limited because direct measurements of temperature have been made for only a short period, approximately a century. Paleoclimatic records from tree rings and corals can cover a longer period, but the data have problems of

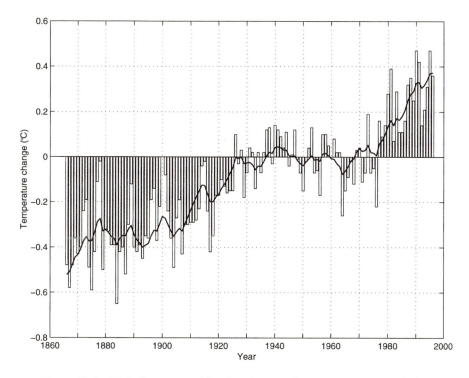

Figure 13.4 Globally averaged land and sea surface temperatures, relative to the average for the period 1951 to 1980, over the past century. Because of the paucity and low quality of the data, especially in the nineteenth century, the early part of the record has considerable uncertainty.

interpretation. Some significant climate fluctuation—for example, the "Little Ice Age" of several centuries ago, described in chapter 11—apparently occurred in the absence of any changes in atmospheric carbon dioxide levels (see fig. 13.2). Such climate fluctuations could therefore be part of our planet's natural climate variability. In that case, detection of the current anthropogenically induced signal will be difficult until it is very large.

What the Models Predict

Temperature fluctuations in a small region—the neighborhood of Chicago say—can be enormous over a short period of time, a day or two, for example. Because they are typically associated with the movements of air masses, those fluctuations, tend to cancel if we take averages over a very large area and over prolonged periods. Spatial and temporal averages can similarly decrease the uncertainties in the results from climate models. Predictions have the least uncertainty when they are for large areas and prolonged periods, and the greatest uncertainty when they are for a small region and a short period of a few weeks or months. The following results are from models in which the atmospheric concentration of greenhouse gases continually rises at the present rate.

Very Probable. The increase in globally averaged surface temperatures over the period 1990 to 2050 will be in the range of 0.5 to 2°C. Globally averaged precipitation will increase; arctic land areas will experience wintertime warming; global sea level will rise by 5 to 40 cm by the year 2050.

Probable. Rainfall will increase over the high latitudes of the northern hemisphere, but will decrease in the midlatitudes of northern hemisphere continents. This suggests that it will be possible to extend farming farther north in Canada and Siberia.

Uncertain. Climate changes in relatively small regions, with dimensions on the order of a few thousand kilometers, will be different from the globally averaged changes. There will also be changes in the frequency and intensity of phenomena such as hurricanes, floods, prolonged droughts, and El Niño.

The various models all predict that warming will be far more pronounced in high northern than in high southern latitudes. One reason for this asymmetry relative to the equator is the distribution of continents that allows a Circumpolar Current in the southern oceans but not the northern ones. The Antarctic Circumpolar Current blocks poleward currents that can transport warm water to high southern latitudes. The absence of such a circumpolar current in the north permits poleward currents—the Gulf Stream, for example—to transport large amounts of warm water to high northern latitudes. This asymmetry is fortuitous because, if a warming of Antarctica should cause the enormous glaciers over that continent to melt, then the associated rise in sea level would be considerable. Melting of the ice over the polar ocean would have a minimal effect on sea level because that ice is already floating on water, unlike the ice over Antarctica, a continent.

Warming of high northern latitudes may not cause a huge rise in sea level but can have other disturbing consequences. Recall that the deep waters of the oceans are rich in carbon dioxide because they are cold; they will lose that gas to the atmosphere should their temperatures increase. The thermohaline circulation of the ocean, discussed in chapter 8, maintains the low temperatures of the deep ocean. That circulation involves the sinking of dense, saline water in the northern Atlantic Ocean. A significant rise of temperatures in that region, and an associated increase in rainfall that decreases the salinity of the surface waters of the northern Atlantic, could make those surface waters so buoyant that they no longer sink. In that case, the thermohaline circulation would be altered. The consequences could include accelerated global warming.

Although many people are not alarmed by a warning that global temperatures will rise 2°C over several decades, they do become concerned about the prospect of regional climate changes such as an increase in the occurrence of severe storms and floods, or prolonged droughts, with an adverse effect on agriculture and other human activities, and the spread of certain diseases into regions where they are uncommon at present. These aspects of global warming, which have the largest uncertainties, are becoming the focus of research efforts.

Epilogue

The industrial and agricultural activities that maintain our standard of living are causing a rapid increase in the atmospheric concentration of greenhouse gases. Up to now we have been gambling that the benefits will far outweigh any possible adverse consequences. Some experts are warning us that we are making poor bets, that global warming is already underway, but others assure us that global warming is only a remote possibility that can safely be ignored. Whom should we believe?

Experienced gamblers are not surprised when the experts disagree; they realize that emotional and ideological factors influence the evaluation of uncertain information. They therefore familiarize themselves with the available information concerning Earth's habitability and the sensitivity of our planet's climate to perturbations. They quickly discover that Earth is immensely complex and that failure to recognize this fact readily leads to controversies. For example, is our planet fragile or robust? It is both! On the one hand, Earth has always been hospitable to life, in spite of a large increase in the intensity of sunlight over the past four billion years; from the perspective of the biosphere as a whole, our planet is robust (see chap. 10.) On the other hand, very modest changes in the distribution of sunlight on Earth over the past million years have caused dramatic climate changes such as recurrent ice ages; to an individual species, Earth can appear fragile (see chap. 11). Our current actions are cause for concern because of the extent to which we are interfering with the processes that make this a habitable planet. We are increasing the atmospheric concentrations of several greenhouse gases, not by a small percentage, but by factors of two and more. Particularly disquieting is the rapid rate of increase; the growth is exponential, a dangerous situation that calls for action long before there is clear evidence of impending trouble. Recall the example of exponential growth in chapter 1: a pond in which the number of lilies doubles each day until they fill the entire pond on day 100. Can the gardener afford to defer action until, say, one eighth of the pond is filled with lilies? In that case, he will wait until day 97 to prevent a disaster on day 100! The experts who disagree about global warming are in effect debating whether we are at day 20, 40, or 60. Though resolution of the arguments is an important matter, far more important is recognition of the explosive nature of the problem, its exponential growth. In coping with problems of that type it is wise to act sooner rather than later.

The computer models of Earth's climate that provide quantitative estimates of future global warming are continually being refined so that the uncertainties in their results are decreasing steadily, causing the spectrum of scientific opinions to converge slowly. Nonetheless, the scientific results will always have uncertainties (because weather and climate are immensely complex phenomena), and some scientists will always dissent. (Skepticism is essential to progress in science, as explained in chapter 1.) To defer decisions continually in the expectation of precise scientific predictions endorsed unanimously by the experts, is to behave as naively as the South African Boer in the following story.

An elderly South African Boer from a remote part of the country is preparing for his first visit to distant relatives in the big city Pretoria. His children patiently explain the intricacies of the momentous journey to him. First he must travel on a local train to the junction where he changes to the fast train from Cape Town to Pretoria. They repeatedly rehearse with the old man what he has to do at the junction: get off the local train; cross the tracks by means of a bridge; wait for five minutes on the platform opposite the one where he had arrived; and board the fast train to Pretoria. On the big day everything goes smoothly. At the junction, the old Boer executes his maneuvers perfectly and in due course he is seated comfortably on the fast train. It takes a few minutes to recover from the excitement at the junction, but, after a while, the old man relaxes and starts a conversation with the passenger opposite him. He is perplexed to learn that the gentleman is traveling to Cape Town, not Pretoria. The puzzled old man is quiet for a long time but finally he sees the light. With a broad grin he remarks to his fellow passenger that modern science really is a miracle. "This train, for example. You, over there, are on your way to Cape Town while I, facing the opposite direction, am on my way to Pretoria."

APPENDIXES

APPENDIX 1

BETWEEN THE IDEA AND THE REALITY

A1.1. Exponential Growth and Decay

If the number of lilies in a pond doubles each day, then the number is said to grow exponentially. Each number in the sequence is double the previous one.

$$1, 2, 4, 8, 16, 32, 64, 128, 256, 512$$

During the course of a day, the increase ΔN in the number of lilies is proportional to the number of lilies N already in the pond at the start of the day

$$\Delta N = N.$$

In general, exponential growth occurs when the increase ΔN during an interval of time Δt is proportional to N.

$$\Delta N = \lambda N \Delta t,$$

where λ is the rate of growth. If the growth is continuous rather than discrete then, from this equation, it can be shown that the value of N at any time t is given by

$$N = N_0 e^{\lambda t},$$

where N_0 is the initial number (at time $t = 0$) and $e = 2.718282$. If λ has a negative value, then the formula describes exponential decay. Figure A1.1 depicts these functions, which are relevant to a great variety of phenomena including not only the growth of the lilies in the pond, but also the growth of money deposited in a bank at a fixed interest rate λ, and of the human population that currently is growing at approximately 1.8% each year. With an appropriate negative value for λ, the formula also describes the rate at which the density and pressure of the atmosphere decrease with increasing height (see chap. 4).

A useful rule of thumb for phenomena involving exponential growth is the following: the time it takes for the quantity to double is approximately 70 divided by the percentage growth rate. For example, if the human population is growing at 1.8% per year, then it will

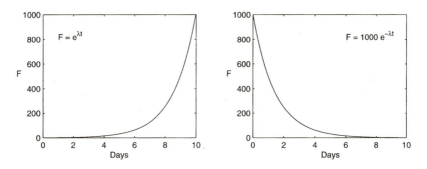

Figure A1.1 *Left*: The function $F = F_0\exp(\lambda t)$ where $F_0 = 1$ and $\lambda = 1/(.693$ days) so that the doubling time for the exponential growth is 1 day. *Right*: The function $F = F_0\exp(-\lambda t)$ where $F_0 = 1000$ and $\lambda = (1/.693$ days) so that the half-life for exponential decay is 1 day.

double in $70/1.8 = 39$ years. At such a rate, there will be more growth during the next four decades than has occurred in all of human history.

A1.2. Establishing a Chronology

Initial estimates of the age of Earth varied considerably, from about 6000 years (on the basis of a literal interpretation of biblical genealogies) to practically infinity. (James Hutton, regarded as the father of modern geology, proposed that the natural processes that are modifying the surface of the Earth today operated uniformly and continuously in the past, declaring there is "no vestige of a beginning, no prospect for an end.") That Earth did have a beginning and that its age greatly exceeds 6000 years became evident once geologists used the progression of life-forms in the fossil record to establish the sequence of eras, periods, and epochs shown in figure A1.2. In this scheme, sedimentary rock "systems" from different parts of the world are assigned to the same period if they have similar fossils. The relative positions of layers of rocks determine their relative ages; the deeper the rock, the older it is. Rocks from the Paleozoic ("old life") Era have the fossils of invertebrates such as mollusks and trilobites, and also of fishes and amphibians. Reptiles, especially dinosaurs, appear next during the Mesozoic ("middle life"). The uppermost rocks from the Cenozoic ("recent life") are rich in fossils of mammals. The eras are divided into periods, some of whose names indicate where geologists first identified them. Very ancient rocks were first studied in Wales, which the Romans called Cambria and which was originally

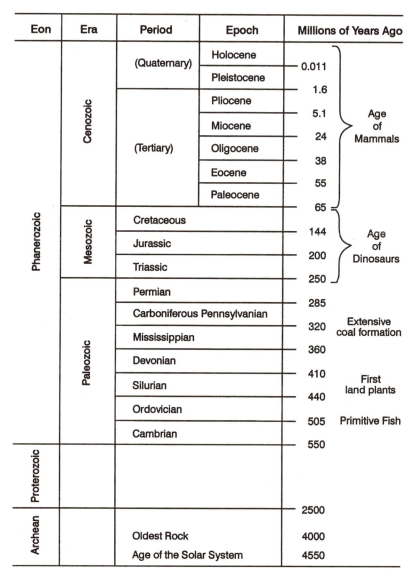

Eon	Era	Period	Epoch	Millions of Years Ago
Phanerozoic	Cenozoic	(Quaternary)	Holocene	0.011
			Pleistocene	1.6
		(Tertiary)	Pliocene	5.1
			Miocene	24
			Oligocene	38
			Eocene	55
			Paleocene	65
	Mesozoic	Cretaceous		144
		Jurassic		200
		Triassic		250
	Paleozoic	Permian		285
		Carboniferous Pennsylvanian		320
		Mississippian		360
		Devonian		410
		Silurian		440
		Ordovician		505
		Cambrian		550
Proterozoic				2500
Archean		Oldest Rock		4000
		Age of the Solar System		4550

Age of Mammals (65–0.011)
Age of Dinosaurs (250–65)
Extensive coal formation
First land plants
Primitive Fish

Figure A1.2   The geologic time scale.

inhabited by two ancient tribes, the Ordovician and the Silures. Perm is a region in Russia. Cretaceous means chalky, referring to deposits of calcium carbonate that have similar fossils. (The terms *Primary* and *Secondary*, which are no longer in use, referred to primitive granitic or metamorphic rocks and to indurated sedimentary rocks. The terms

Tertiary and *Quaternary* are still in use for the more friable and younger rocks.)

An accurate chronology was not possible until early in the twentieth century when scientists developed dating methods that rely on radioactive elements, which change spontaneously into other elements, known as daughter products. These changes occur in the nuclei of atoms. An atom of any element has electrons spinning around a nucleus composed of protons and neutrons. The identity of an element is determined by the number of protons. The number of protons plus neutrons, known as the atomic number of an element, essentially determines the mass of an atom. Different atoms of the same element that have different masses are known as isotopes of that element. In the case of the element carbon C, possible isotopes have 6, 7, and 8 neutrons and are denoted by the symbols ^{12}C, ^{13}C, ^{14}C where the superscript denotes the atomic mass. (The first of these isotopes is the most common form of carbon.) If the nitrogen isotope ^{14}N, the most common isotope in the atmosphere, should absorb a neutron and emit a proton—this happens high in the atmosphere because of the presence of energetic cosmic rays—its atomic mass remains unchanged, but it will be transformed into ^{14}C, an isotope of the element carbon. Some isotopes are unstable, or radioactive, and reach a more stable internal organization by transformations in the nucleus. The isotope ^{14}C is unstable and spontaneously reverts to ^{14}N. We cannot predict exactly when a specific nucleus will undergo this transformation. However, given a large number, N, of nuclei at a certain time, the number ΔN that decays radioactively over a short time interval Δt is directly proportional to N and to Δt. As explained in Appendix A1.1, it follows that the number N decreases exponentially with time t according to the expression,

$$N = N_0 \exp(-\lambda t)$$

where N_0 is the number of nuclei at time $t = 0$. The time over which the population decreases by a factor of two is known as the *half-life* of the decay process. For ^{14}C, the half-life is 5730 years. This result can be used to establish a precise radiocarbon clock for events over the past 40,000 years.

Cosmic rays continually produce the isotope ^{14}C in the atmosphere where it interacts with oxygen to produce $^{14}CO_2$, which chemically behaves like $^{12}CO_2$, ordinary carbon dioxide. The ratio of these two isotopes in the atmosphere remains unchanged with time, provided there is no change in the cosmic ray production of radiocarbon and no change in the supply of $^{12}CO_2$ to the atmosphere. The value of that ratio will initially be shared by various forms of life that interact with the atmospheric pool of carbon dioxide. Once sequestered from the atmosphere, in organic matter, the ratio changes because the $^{14}CO_2$ decays. Hence, by determining the ratio in ancient organic material—the shroud of Turin, for example—we can determine the age of the material because we know the rate of decay of the radiocarbon. (After some 40,000 years, too little radiocarbon is left for accurate measurements.)

To establish the dates of older samples, geologists turn to radioactive elements that decay more slowly than ^{14}C. Particularly useful for determining the age of the solar system (from meteorites on Earth) are isotopes of uranium that decay, with a half-life of about a billion years, to produce daughters of lead. The isotope thorium-230 decays at a rate that makes it suitable for establishing dates up to a few hundred thousand years ago. Naturally occurring elements that are useful in the range of about a million years to a few billion years include potassium-40. For none of these radioactive isotopes do we know the initial amount in the rock. We can, however, measure the amount present at this time and the amount of stable daughter that is present. Given the half-life, it is then possible to determine an age by assuming that the daughter is entirely the consequence of radioactive decay of the parent.

To minimize uncertainties in the results from radioactive elements, scientists rely on a variety of complementary dating methods. One of them depends on abrupt reversals of the polarity of Earth's magnetic field. At present, the needle of a compass points north, but at different times in the past it would have pointed in the opposite direction. (The reasons for the repeated reversals of Earth's magnetic field are not known.) Lava flows acquire a direction of magnetization parallel to Earth's magnetic field at the time of the flow so that ancient lava flows, even if they are on different continents, synchronously record the same reversal events.

Further Reading.   For a discussion of the reasons why Malthus' predictions for England proved wrong, see Kennedy (1993). Articles by Holmes (1994) and by Stone (1994) discuss the New England fisheries crisis of the 1990s. Benedick (1991), the ambassador who represented the United States in negotiations that led to the Montreal Protocol, describes those negotiations in his book *Ozone Diplomacy*. Scientific issues concerning global warming are discussed comprehensively in the reports of the International Panel on Climate Change, edited by Houghton et al. (1990, 1995).

APPENDIX 2

IS OUR PLANET FRAGILE OR ROBUST?

A2.1. Gaia

Our solar system has nine planets, but only one, Earth, has life. Could that be the reason why only Earth has conditions that favor life? In other words, does life itself create the conditions it requires? The Gaia hypothesis states that the biosphere, the assemblage of all life-forms on Earth, acts as an organism that maintains conditions that are favorable to life. Even if true, the idea is not particularly reassuring to any one species because Gaia does not hesitate to sacrifice some species should it benefit the whole. (Dinosaurs are no longer with us.) *Homo sapiens*, late in the twentieth century, has evolved to a stage where it is very vulnerable to even small changes in climatic conditions. Judging by the geological records, Gaia tolerates huge changes that could be catastrophic for *Homo sapiens*.

If the Gaia hypothesis were simply a statement that life-forms do not merely evolve passively in response to a changing environment but are capable of changing that environment, then there would be no controversy. After all, Earth's atmosphere acquired oxygen only after life evolved. The chemical composition of the atmosphere would change radically should all forms of life suddenly disappear. There is no doubt that life on Earth influences the climate of this planet and that physical conditions in turn affect life. Whether the biosphere is capable of controlling the global environment to its own benefit, whether it is capable of acting, for example, like a thermostat that regulates temperature fluctuations to suit itself, is another matter altogether.

A hypothetical world inhabited by white daisies is often cited as an illustration of how Gaia operates. Suppose that, in this world, the intensity of sunshine increases. Daisies can then keep temperatures stable by increasing their numbers because, the larger the number of daisies, the greater the amount of sunshine that is reflected. Should the Sun grow fainter, a decrease in the number of daisies will prevent temperatures from falling.

This argument assumes that it is desirable that temperatures be stable. For whom? From the perspective of the daisies, the higher the temperature, the better for them, because their numbers increase. Suppose there is an optimal temperature for the daisies so that an increase in temperature beyond that limit causes their number to decrease. When that happens, daisy-world absorbs more sunlight, and temperatures continue to rise, until there are ultimately no daisies. This result can be generalized: if the biota regulate atmospheric conditions, then those conditions cannot be optimal for the biota.

The statement that Gaia maintains optimal conditions for life is problematic because it is difficult to avoid inconsistencies and tautologies in defining what is meant by optimal conditions. Gaia is best regarded as a metaphor, not as a testable scientific hypothesis. For a discussion of this topic, see Lovelock (1988) and Kirchner (1989).

A2.2. Chaos

The motion of a wallet sliding down a ski slope with moguls is not entirely chaotic. If followed over a considerable distance, it becomes evident that the wallet has a tendency to move in either a southeasterly or southwesterly direction as is evident in figure A2.1 The motion is said to have a "strange attractor" whose properties are usually depicted on a phase diagram that has as axes the west-east and north-south velocity components. For an excellent discussion of this and other issues related to chaos, see Lorenz (1993).

Further Reading.   Recent textbooks that cover various aspects of the geosciences and Earth's habitability include those of Broecker (1985), Graedel and Crutzen (1993), Trenberth (1992), and Turco (1997).

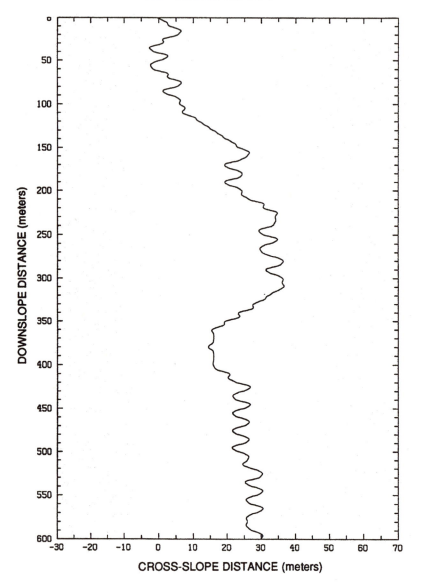

Figure A2.1 The path of a wallet down a ski slope with moguls. From Lorenz (1993).

APPENDIX 3

LIGHT AND AIR

A3.1. Earthshine

The albedo of Earth, the fraction of the incident sunlight that it re-flects, can be measured accurately by means of satellites, or by photo-graphing the new moon. Sometimes, when we look at the new moon for a prolonged period, we get the impression that we see, not just a sliver in the sky, but the entire circular disk. This is no illusion. Photo-graphs taken with a sufficiently long exposure do provide images of the entire disk. The sliver is lit directly by sunlight; the rest of the disk is lit indirectly by sunlight after it has been reflected by Earth. Photo-graphs therefore capture sunlight that is first reflected by Earth—it is sometimes called earthshine—and that is subsequently reflected by the Moon. Such photographs provide a measure of Earth's albedo.

A3.2. The Scattering of Light

The atmosphere has molecules of various gases plus particles of dust, ice, etc. of various sizes. The extent to which such particles—assumed to be perfect spheres of radius r—scatter light depends on the relative magnitudes of r and λ, the wavelength of the light. It is therefore useful to introduce a parameter that measures this ratio:

$$\alpha = 2\pi r / \lambda.$$

Let S be a measure of the extent to which a particle scatters light; then, for particles so small that the value of α is much less than one,

$$S \sim \alpha^4.$$

This result, which describes Rayleigh scattering (in honor of the scientist who first studied this phenomenon), can be used to deter-mine the relative effects of an assemblage of small particles (or mole-cules) on blue and red light. We expect blue light ($\lambda = .47\ \mu m$) to be scattered more readily than red light ($\lambda = .64\ \mu m$) because it has a shorter wavelength. In quantitative terms, the difference, which is the principal reason why the sky is blue, is

$$S = (.64/.47)^4 = 3.45.$$

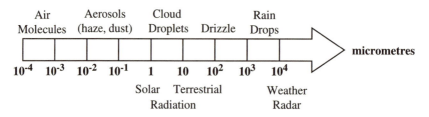

Figure A3.1   Relative sizes of various particles and of different wavelengths.

If the particles are extremely large so that the value of α exceeds 50, approximately, then scattering is practically independent of wavelength. Hence, if their particles are large, pollutants in the atmosphere cause the atmosphere to be whitish in appearance.

Figure A3.1 shows the relative magnitudes of different particles and different wavelengths. Note that weather radar uses very long wavelengths that can be reflected by large water droplets in clouds that precipitate.

A3.3. Blackbody Radiation

The radiation E of a blackbody, one that absorbs all the radiation incident on it, covers a spectrum of wavelengths, λ, and depends on the temperature T of the surface (measured in degrees Kelvin), as shown in figure A3.2:

$$E = \frac{c_1}{\lambda^5 \, [\exp(c_2/\lambda T) - 1]}, \tag{A3.1}$$

where

$$C_1 = 3.74 \times 10^8 \text{ W m}^{-2}\mu\text{m}^4$$
$$C_2 = 1.44 \times 10^4 \text{ } \mu\text{mK}$$

The radiation has a maximum value at a wavelength

$$\lambda_{\max} = a/T \quad \text{where } a = 2897 \text{ } \mu\text{mK}. \tag{A3.2}$$

The total emission Q^*, integrated over all the wavelengths, is given by the formula, discovered toward the end of the nineteenth century by the physicists Stefan and Boltzman,

$$Q^* = \sigma T^4, \tag{A3.3}$$

where σ is a constant with the value 5.67×10^{-8} watts/(m^2K^4).

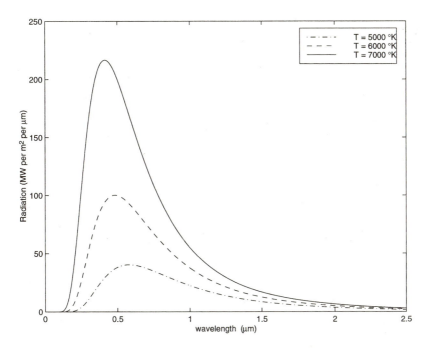

Figure A3.2   Radiation from surfaces at different temperatures.

A3.4. Effective Temperatures of the Planets

The Sun is the principal source of energy for the planets. On Earth, at a distance D_e from the Sun, we measure the intensity of sunlight to be 1372 watts per square meter. This amount, Q_e, falls on a surface that is perpendicular to the Sun's rays and that has an area of 1 square meter. Given this information, the distances of the planets from the Sun, and certain laws of physics, it is possible to calculate the temperature of the Sun and of each of the planets. If a planet has an atmosphere that provides a greenhouse effect, then this effective temperature, as it is known, is the temperature in the upper layers of the atmosphere. The difference between the temperature at the surface and that aloft provides an estimate of the greenhouse effect.

The Sun radiates heat of intensity Q_e across the surface of a sphere of radius D_e. Therefore,

$$\text{Total radiation from the Sun} = 4\pi D_e^2 Q_e. \qquad (A3.4)$$

Now suppose we move to another planet, at a distance D_p from the Sun, where the intensity of solar radiation is Q_p. It follows from similar arguments that

$$\text{Total radiation from the Sun} = 4\pi D_p^2 Q_p. \qquad \text{(A3.5)}$$

These two expressions must be identical (if interstellar space absorbs no sunlight) so that the radiation that any planet receives is

$$Q_p = Q_e D_e^2 / D_p^2. \qquad \text{(A3.6)}$$

The continual absorption of sunlight would lead to a steady increase in temperature, if there were not a loss of heat to offset the gain. Every object loses heat to its surroundings at a rate that depends on its temperature. Hence, if the temperature of a planet were to increase because of the absorption of heat from the Sun, then the rate at which it radiates heat to space will also increase. In due course, there will be a balance between the rates at which heat is absorbed and lost. In such a state of equilibrium, the temperature of the planet remains constant. In calculating how much heat a planet absorbs, we must keep in mind that in low latitudes the planet's surface is generally perpendicular to the Sun's rays, but that is not the case in high latitudes. We can circumvent this difficulty by recognizing that, seen from the Sun, the planet is a disk that intercepts its rays. The area of that disk is πR^2, where R is the radius of the planet. Hence, the planet absorbs a total of $\pi R^2 Q_p$ watts. Not all the incident sunlight is absorbed, however; a fraction that depends on the albedo α is reflected. Therefore,

$$\text{Heat absorbed} = (1 - \alpha)\pi R^2 Q_p. \qquad \text{(A3.7)}$$

In a state of equilibrium, the planet must radiate as much heat as it absorbs. The entire surface of the spherical planet, which has an area $4\pi R^2$, radiates. If the planet radiates Q^* watts/m^2, then

$$4\pi R^2 \, Q^* = (1 - \alpha)\pi R^2 Q_p \qquad \text{(A3.8)}$$

so that

$$Q^* = (1 - \alpha)Q_p/4,$$

a result that is independent of the size (or radius) of the planet.

It is now possible to calculate the temperature of a planet from the Stefan and Boltzman formula (A3.3) which is accurate even for bodies that are not entirely black—the Sun and other stars, for example.

$$\begin{aligned} T_e &= [(1 - \alpha)Q^*/4\sigma]^{.25} \\ &= [(1 - \alpha)Q_e D_e^2 / 4\sigma D_p^2 \,]^{.25}. \end{aligned} \qquad \text{(A3.9)}$$

TABLE A3.1
Temperatures of the Planets

Planet	Distance from Sun (10^6 km)	Insolation (watts/m²)	Albedo	T_s (°C)	T^* (°C)	T_e (°C)
Mars	228	589	.15	−53	−47	−56
Earth	150	1372	.30	+15	+6	−18
Venus	108	2613	.75	+430	+55	−41

Note: The third column (insolation) shows the intensity of the sunlight that reaches each planet; the next column indicates the fraction of the incident sunlight that is reflected. T_s is the globally averaged temperature at the surface of each planet; T^*, the temperatures they would have if they absorbed all the incident sunlight and had no atmospheres; and T_e, is the same as T^* except that the nonzero albedo is taken into account.

This formula enables us to calculate the effective temperature of any planet, provided we know the intensity of solar radiation at Earth, the distance of Earth from the Sun, and the distance of the planet under consideration from the Sun. Table A3.1 gives numerical values that are shown graphically in figure 3.1.

A3.5. The Greenhouse Effect

A highly idealized model of the atmosphere that illustrates its greenhouse effect is shown in figure A3.3, in which the atmosphere is simply a thin sheet of glass of uniform temperature. Rays from the Sun penetrate to the surface of the planet, which therefore warms up. As temperatures rise, the surface radiates more and more infrared heat upward, heat that the atmosphere, the sheet of glass, absorbs. The glass therefore warms up and emits radiation, from both its upper and lower surfaces, down to the floor and up to space. The radiation from the upper surface of the glass is what the planet as a whole loses and, in a state of equilibrium, that loss must equal the heat the planet as a whole gains. As figure A3.3 shows, if the planet gains S units of sunlight (which the ground absorbs), then the upper surface of the glass must be radiating S units. This, once again, is an application of the law for the conservation of radiative heat to the planet as a whole.

The radiation emitted by any surface depends on its temperature. If we assume that the glass is thin so that it has a uniform temperature, then it follows that the upper and lower surfaces of the glass have the same temperature, and hence emit the same radiation S. That radiation is upward in the case of the upper surface, downward toward the floor in the case of the lower surface.

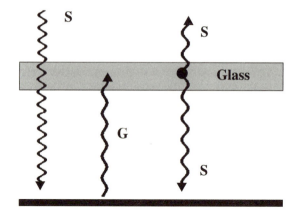

Figure A3.3 An idealization of the greenhouse effect in which the greenhouse glass is entirely transparent to sunlight, of intensity S, from above, and is entirely opaque to infrared heat, of intensity G, from the surface below. In a state of equilibrium the glass attains a temperature which is such that it radiates S units from each of its surfaces.

The surface of the planet therefore has two sources of heat; it absorbs S units from the glass plus S units from the Sun. Because of the presence of the glass, the surface receives twice as much radiation as it would otherwise have received! It therefore has a higher temperature than it would have had in the absence of the glass. Note that these results are entirely consistent with the law for the conservation of radiative heat. The glass absorbs as much heat (G) as it radiates (2S). The same is true of the surface of the planet.

The surface is at a higher temperature than the glass because the one radiates 2S units, the other S units. This means that temperatures decrease with elevation. (It is understandable that temperatures should be higher beneath the blanket than above it.) This result sheds new light on the discrepancy between the calculations and measurements in figure 3.1. Recall that the calculations did not take planetary atmospheres into consideration and therefore are relevant to the surfaces of planets without atmospheres. In the case of a planet that has an atmosphere, the calculations remain relevant, not to the surface, but to the cooler upper atmosphere. The difference between the temperatures of the surface and of the upper atmosphere is therefore a measure of the greenhouse effect.

The realism of the very idealized model in figure A3.3 can be improved by means of simple refinements that depend on the planet under consideration. To simulate Venus, the one layer of glass that absorbs all the incident infrared radiation is inadequate; that layer provides a greenhouse effect too small to explain the high tempera-

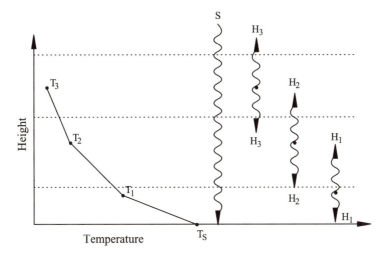

Figure A3.4   A model that divides the atmosphere into three layers. See the text for definitions of symbols.

tures on Venus. The model needs several layers of glass to be relevant to Venus, a situation dealt with in the next exercise. In the case of Earth, one layer is more than enough because Earth's atmosphere absorbs, not all the infrared radiation from the surface as in figure A3.3, but only a fraction of that radiation. The model can be refined further by distributing the atmospheric gases vertically as observed instead of confining them to a thin layer as in figure A3.3.

The gases in the atmosphere are not conveniently confined to a thin layer corresponding to a sheet of glass. A better approximation would be to divide the atmosphere into several layers, one on top of the other. In figure A3.4, the atmosphere is divided into three layers, all transparent to sunlight that penetrates to the surface, which is heated to a temperature T_s. The surface radiates heat G, which, we shall assume, is absorbed entirely by the first layer so that its temperature increases to T_1. This layer radiates heat H_1 upward and an equal amount downward, toward Earth's surface. The heat H_1 that is radiated upward is absorbed entirely by the second layer, which radiates heat H_2 upward and downward. The third layer receives heat only from the second layer and maintains a temperature T_3 by radiating heat H_3 upward and an equal amount downward. In a state of equilibrium, Earth gains as much heat from the Sun as it radiates to space so that

$$H_3 = S.$$

The third layer receives an amount H_2 and radiates $2H_3$:

$$H_2 = 2H_3 = 2S.$$

For the second layer,

$$H_1 + H_3 = 2H_2 \text{ so that } H_1 = 3S.$$

For the ground,

$$S + H_1 = G \text{ so that } G = 4S.$$

It is now possible to calculate the temperature of the surface and of each layer from the Stefan-Boltzman equation:

$$T_s = (4S/\sigma)^{1/4}$$
$$T_1 = (3S/\sigma)^{1/4}$$
$$T_2 = (2S/\sigma)^{1/4}$$
$$T_3 = (S/\sigma)^{1/4}.$$

Temperatures therefore decrease with height as shown in figure A3.4.

The crucial assumption in this calculation is that each layer exchanges heat only with its immediate neighbors. In the figure, the thickness of a layer increases with increasing height because the density of an absorbing atmospheric gas decreases with height so that radiation travels farther before being absorbed. If a layer is too thin, then the radiation will traverse it; if it is too thick, then the radiation that the layer emits will be reabsorbed within the layer. The total number of layers into which an atmosphere can be divided in this manner is known as its optical thickness. It is a simple matter to extend our results for a three-layer atmosphere to one of arbitrary optical thickness (with an arbitrary number of layers). In that case,

$$\sigma T_s^4 = (1 + \text{Optical thickness}) \, S.$$

In other words, the temperature of the surface increases with an increase in the optical thickness, which depends on the efficiency with which atmospheric gases absorb infrared heat radiation. Increasing the concentration of greenhouse gases in our atmosphere will increase its optical depth and hence the surface temperature. At present, the optical thickness is such that the surface is at a temperature of $+15°C$, rather than $-18°C$, the temperature the surface would have if there were no atmosphere.

Exercises.

1. Calculate the temperature of the Sun from equation (A3.2) given that the maximum solar emission is at about 0.475 μm, which corresponds to blue light. (The Sun appears more yellow than blue because most of the radiation is emitted at wavelengths longer than that of the peak monochromatic irradiance.)

2. Explain why the colors of stars are related to their temperatures but the colors of planets are not.

3. What is the temperature of the Moon, given that the intensity of sunlight there is the same as it is on Earth?

4. Calculate the temperature of the Sun, given that its radius is 7×10^5 km and that, at our distance from the Sun, 150×10^6 km, the intensity of sunshine is 1372 watts/m². (The surface area of a sphere of radius R is $4\pi R^2$.)

5. Because Earth's orbit is an ellipse, its distance from the Sun varies by about 3%. (It is closest in early January, farthest in July.) Show that the corresponding seasonal change in Earth's effective temperature is about 4° Kelvin.

6. A new greenhouse glass on the market is transparent to sunlight and absorbs 80% of the infrared radiation that the ground emits. If the intensity of the incident sunlight is S units, show that, because of the glass, the floor of a greenhouse will receive $5S/3$ units.

7. Explain why an overcast sky favors lower temperatures during the day but higher temperatures at night.

Further Reading.   Gribben (1984) provides the layman with an excellent introduction to quantum mechanics. The fine text by Wallace and Hobbs (1977), which assumes a knowledge of calculus, covers many of the topics in this and subsequent chapters.

APPENDIX 4

WHY THE PEAK OF A MOUNTAIN IS COLD

A4.1. The Scale Height of the Atmosphere

Gravity keeps most of the air molecules near Earth's surface so that density and pressure decrease rapidly with height. Although Earth's gravitational attraction decreases with increasing distance from the surface of Earth, this is not a factor that influences the height of the atmosphere significantly. (A passenger in a commercial jet at a height of 10 km has most of the mass of the atmosphere below him but is not aware of a decrease in the force of gravity; he is not aware of any detectable change in his weight.) To calculate the height of the atmosphere, we therefore assume that gravity is a constant.

The ideal gas equation can be written

$$p = RDT, \qquad (A4.1)$$

where p, D, and T are the pressure, density, and temperature (in degrees Kelvin) of a unit mass (1 kg) of the gas, and R is a constant whose value depends on the particular gas under consideration. In an isothermal atmosphere (T = constant), this equation becomes Boyle's law, which states that

$$p = aD, \qquad (A4.2)$$

where the constant a has the value RT.

The second relation between p and D is the hydrostatic law, which states that the pressure difference Δp between two points separated by a vertical distance h is

$$
\begin{aligned}
\Delta p &= \text{weight of a column of air of height } h \\
&= g \times \text{mass of a column of air of height } h \\
&= g \times D \times \text{volume of a column of air of height } h \\
&= g \times D \times h
\end{aligned}
$$

because the column covers a unit area. Here g is gravity and D is the density of the air. We have assumed that the density D is constant. In reality, it decreases with height so that this equation is valid only if the height h is small. It forces us to divide the atmosphere into very thin layers (or pancakes), one stacked on top of another, each with a constant density that is slightly less than that of the layer below and

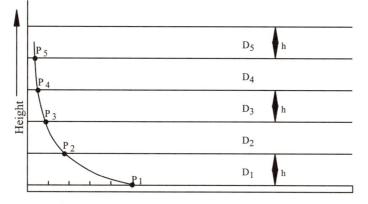

Figure A4.1 The atmosphere divided into layers of depth h, and of density D_1, D_2, etc. The pressures at the bottom of each layer are P_1, P_2, etc.

each of height h (see fig. A4.1). The lowest layer has the highest density D_1, the next layer a smaller density D_2, and so on. If the air is motionless, then the pressure at the ground P_1, is simply the pressure P_2 at height h plus the weight per unit area of the first layer of air,

$$P_1 = P_2 + ghD_1.$$

Similarly,

$$P_2 = P_3 + ghD_2$$
$$P_3 = P_4 + ghD_3. \tag{A4.3}$$

Let us next apply equation (A4.2) to each layer. The average pressure for the first layer is $\frac{1}{2}(P_1 + P_2)$ so that equation (A4.2) becomes

$$\frac{1}{2}(P_1 + P_2) = aD_1$$
$$\frac{1}{2}(P_2 + P_3) = aD_2 \tag{A4.4}$$
$$\frac{1}{2}(P_3 + P_4) = aD_3,$$

and so on. If we now combine equation (A4.3) for difference in pressure between adjacent layers and equation (A4.4) for the average pressure in each layer, then we obtain

$$P_2 = \frac{2-b}{2+b}P_1$$

$$P_3 = \left(\frac{2-b}{2+b}\right)^2 P_1$$

$$P_4 = \left(\frac{2-b}{2+b}\right)^3 P_1,$$

and so on, where

$$b = gh/RT.$$

We chose h to be the height of each layer, but we did not specify what the height is. Let us choose h to equal H, where

$$H = RT/g, \tag{A4.5}$$

then $b = 1$, and we find that the pressure falls by a factor of 3 when the height increases by an amount H. A more accurate derivation of these expressions shows that

$$p = P_1 e^{-z/H}, \tag{A4.6}$$

where z measures distance upward from the surface and $e = 2.718$. This means that the pressure falls by a factor e when the height increases by H. The density decreases similarly with elevation.

For dry air without water vapor, the constant R has the value $R = 287$ J kg^{-1} deg^{-1}. To take the effects of moisture into account, in equation (A4.5), we must replace the temperature T by the virtual temperature T_v, which is defined as the temperature of dry air having the same density and pressure as a given sample of humid air. T_v depends on the relative humidity of the air (discussed in the next appendix) and is higher than the actual temperature, but seldom by more than 3°K. For a mixture of gases that corresponds to those in the troposphere, the scale height is approximately 10 km.

A4.2. The Adiabatic Lapse Rate

Sunshine penetrates through the atmosphere so that it is heated from below. Convection redistributes heat and, if that were the only process affecting the thermal structure of the atmosphere, then temperatures would decrease with height at the adiabatic lapse rate. To determine that lapse rate, we proceed as follows.

The addition of heat to a substance generally increases its temperature unless the substance changes phase—from ice into water, for example. The amount of heat required to change the temperature of 1 g of the substance by 1°C is defined as the specific heat of that substance. Of the naturally occurring substances, water has one of the highest specific heats: 1 cal g^{-1} C^{-1}. Hence, 1 calorie raises the temperature of 1 g of water by 1°C. From the first law of thermodynamics,

$$\begin{aligned} \text{Energy absorbed} &= \text{Change in internal energy} \\ &= mC\Delta T, \end{aligned} \tag{A4.7}$$

where m is the mass of the substance, C is its specific heat, and ΔT is the increase in temperature.

The addition of an amount of heat ΔQ to a unit mass of air causes the temperature of the air to increase by an amount ΔT; it also causes the parcel of air to expand, thus doing work on the surroundings by pushing it away. Hence, the previous equation needs to be modified as follows:

$$\Delta Q = C_v \Delta T + \text{Work Done}, \tag{A4.8}$$

where C_v, the specific heat of the air at constant volume, has the value $718 \text{ J kg}^{-1}\text{K}^{-1}$.

The work done can be visualized by considering air contained in a cylinder fitted with a moveable, frictionless piston (fig. A4.2). If the air expands so that the piston moves a distance L from X to Y, then the work done (ΔW) equals the force exerted on the piston multiplied by the distance L through which the piston moves. The force is the pressure p exerted on the area A of the piston so that

$$\Delta W = pAL = p\Delta V, \tag{A4.9}$$

where ΔV is the increase in the volume of the air. Equation (A4.8) can now be written

$$\Delta Q = C_v \Delta T + p\Delta V. \tag{A4.10}$$

If a process is adiabatic, meaning that no heat is added ($\Delta Q = 0$), then expansion, an increase in the volume, is accompanied by a fall in temperature. Our goal is to determine how much the temperature of a parcel of air falls when it expands because its height in the atmosphere is increased. We need more than the first law of thermodynamics to answer this question. We also have to appeal to the perfect gas law,

$$pV = RT, \tag{A4.11}$$

where R is the gas constant. If the temperature of a parcel changes by an amount ΔT then, from (A4.11), the associated changes in pressure Δp and in volume ΔV are related as follows:

$$p\Delta V + V\Delta p = R\Delta T. \tag{A4.12}$$

If this equation is combined with equation (A4.10) for an adiabatic process, then

$$C_p\Delta T = V\Delta p, \tag{A4.13}$$

where $C_p = C_v + R$ is the specific heat of the gas at constant pressure. This equation relates the decrease in temperature with altitude to the decrease in pressure.

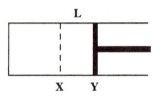

L

X   Y

Figure A4.2   A piston that compresses air by moving a distance L from Y to X.

Finally, we must return to the hydrostatic approximation. The decrease in pressure when the height changes by h is simply equal to the weight of the layer of air of thickness h. The mass of the air is its volume, h, if we consider a unit area, times the density D. To obtain the weight, multiply by gravity g so that

$$\Delta p = Dg\, h.$$

Recall that density D is the inverse of specific volume V so that $DV = 1$. Then $\Delta T/h = G$, where

$$G = -g/C_p.$$

Over a vertical distance h, the temperature changes by g/C_p, a quantity known as the adiabatic lapse rate of temperature. It amounts to a 1°C drop in temperature for a rise of 100 m. If the atmosphere had no source of heat—no gases to absorb earthlight and sunlight, and no release of latent heat—then sunlight would heat Earth's surface and, in a state of equilibrium, temperatures would decrease upward at rate G. (These arguments neglect the effects of water vapor.)

Exercises.

1. The specific heat of sand is only one fifth as large as that of water. Use this fact to explain why the dry sand on a beach in summer feels hot relative to the wet sand.

2. If the pressure at Earth's surface is 1000 mb, at what elevations (in kilometers) will the pressure be 100, 10, and 1 mb? Assume that the atmosphere is isothermal and has a temperature of −20°C.

3. Explain why air released from a tire or a spray can cools its surroundings.

4. Explain why airplanes frequently encounter turbulence immediately after takeoff and before landing on clear, sunny afternoons but not at night.

Further Reading.   For a more complete discussion of the topics of this and subsequent chapters see the texts by Wallace and Hobbs (1977), Ahrens (1994), and Moran and Morgan (1991).

APPENDIX 5

CAPRICIOUS CLOUDS

A5.1. Measuring Moisture in the Atmosphere

Ways to measure the amount of moisture in the atmosphere fall into two groups: the absolute and the relative measures. (In the latter group, the temperature of the air is a factor.)

Water vapor pressure, literally the pressure exerted strictly by the molecules of water vapor in the air, is an absolute measure usually denoted by the letter e. As the amount of water vapor in the atmosphere increases, the vapor pressure e increases, but only up to a certain limit that is reached when the air becomes saturated so that condensation occurs. This saturation pressure, e_s, depends on the temperature of the air and hence is a relative measure of atmospheric moisture. The relationship between e_s and temperature, T, expressed in degrees Kelvin has the approximate form

$$e_s = e_0 \exp[\frac{L}{R}(\frac{1}{T} - \frac{1}{T_0})],$$

where $e_0 = .611$ kPa and $T_0 = 273°K$ are constants; $R = 461$ J^{-1} $K^{-1}kg^{-1}$ is the gas constant for water vapor; and L, the latent heat of vaporization, has the value 2.5×10^6 J kg^{-1}. This expression is an approximation to the Clausius-Clapeyron equation plotted in figure 5.3.

The dew point is the temperature at which the moisture in the atmosphere condenses when the atmosphere is cooled at constant pressure. In figure 5.3, this corresponds to movement from A to C. As the temperature of the air changes, the dew point does not have to change (but the saturation vapor pressure does). Dew point is therefore an absolute measure of moisture.

Relative humidity is the most common measure of atmospheric moisture because it influences our level of comfort. It is a relative measure, because it depends on temperature and is the ratio of the amount of water vapor present in the atmosphere to the maximum amount possible at the current temperature. An equivalent definition is the ratio of the pressure exerted by the water vapor to the maximum pressure that water vapor can exert at the current temperature, expressed as a percentage,

$$RH = \frac{e}{e_s} \times 100.$$

During the course of a day, with no change in the amount of water vapor in the atmosphere, the relative humidity can change from high values at night, when temperatures are low, to low values during the day when temperatures are high. Thus, the presence of fog early in the morning, and its disappearance later in the day as temperatures rise, need not be associated with any change in the amount of water vapor in the atmosphere.

A5.2. Earth's Energy Budget

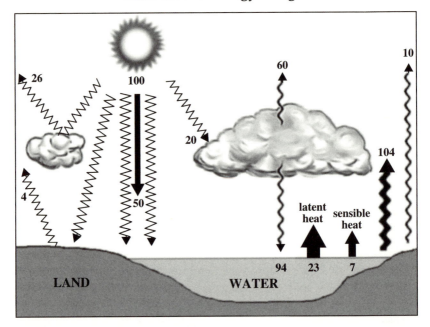

Figure A5.1 Earth's energy budget. Of the solar radiation that Earth receives, 30% is reflected back to space: 4% by land and ice surfaces, 26% by clouds and aerosols. Water vapor and other gases, especially ozone in the stratosphere, absorb another 20% of the incident sunlight so that 50% of it is absorbed at Earth's surface. The greenhouse effect augments these 50 units by 94 units. In a state of equilibrium, the surface must therefore lose 50 + 94 units of heat: 23 through evaporation, 7 through conduction, and 114 through infrared radiation, of which 104 units are absorbed by greenhouse gases in the atmosphere, and 10 are lost directly to space.

A5.3. How Many of Your Molecules Have Been to the Moon?

You participate in the hydrological cycle by consuming, on average, 3 liters of water per day.

1. Assume that your weight is 75 kg and that you are 70% water. What is the shortest time (in principle) that it would take you to replace all the water molecules in your body?

2. Assume that your body exchanges water with a global reservoir of 4×10^{20} kg water that is composed of all the water on Earth (other than that in the deep part of the oceans and in glaciers). The global reservoir includes the molecules that visited the Moon in the bodies of the 18 astronauts that set foot there. Assume that they all visited the Moon in 1970. Estimate the number of molecules that have been in your body and that have also been to the Moon. (Useful facts: 1 liter of water weighs 1 kg and has 332.4×10^{23} molecules.)

Exercises.

1. Explain why, in cold climates, the air indoors tends to be extremely dry.

2. Explain why the globally averaged radiation at Earth's surface is downward.

3. Calculate the upward flux of water vapor, and of the associated energy, at Earth's surface given that the rainfall on that surface is approximately 1 m per year.

Further Reading.   Emanuel (1994) provides a comprehensive discussion of atmospheric convection. For a discussion of Earth's energy budget see the article by Kiehl and Trenberth (1997) and the book by Peixoto and Oort (1992).

APPENDIX 6

THE CLIMATE TAPESTRY

A6.1. Conservation of Angular Momentum

Earth rotates about its axis and hence completes an angle of 2π once a day, or every 8.64×10^4 seconds. We define Earth's angular velocity as

$$\Omega = 2\pi/(8.64 \times 10^4 \text{ sec})$$
$$= 7.3 \times 10^{-5} \text{ sec}^{-1}.$$

The speed of a rotating parcel is its angular velocity multiplied by its distance from the axis of rotation. At the equator, it is ΩR because R, the radius of Earth, is also the distance to the axis of rotation at that latitude. At a higher latitude, θ, the speed is smaller; it is $\Omega R\cos\theta$ (fig. A6.1).

Consider a parcel of air moving on Earth's surface from point B at the equator to point A which has latitude θ. At B, it starts with a speed that is higher than that of parcels rotating with Earth at A. During this motion, and in the absence of any east-west forces (e.g., friction), the angular momentum (not the speed) of the parcel remains constant in accord with Newton's laws of motion. The angular momentum of a parcel of 1 g rotating with Earth at latitude θ is $(\Omega R\cos\theta)R\cos\theta$. For the parcel moving from the equator, the distance from the Earth's axis of rotation decreases from R to $R\cos\theta$ during the journey so that its angular velocity must change from Ω to Ω' to satisfy the conservation of angular momentum,

$$\Omega R^2 = \Omega'(R\cos\theta)^2.$$

Its speed at A is therefore $U = \Omega'R\cos\theta = \Omega R/\cos\theta$. Relative to an observer moving with the solid Earth at A, the speed of the parcel is

$$u = \Omega R/\cos\theta - \Omega R\cos\theta$$
$$= U^*\sin^2\theta/\cos\theta,$$

where U^* is the absolute speed of a particle moving with the solid Earth at the equator. Hence, relative to observers rotating with Earth, the parcel appears to be forced eastward as it moves poleward. This apparent force is the Coriolis force.

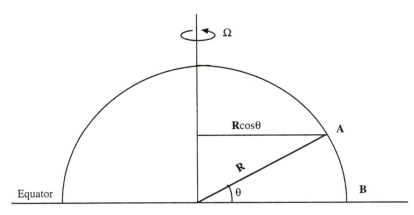

Figure A6.1   A particle that moves from B at the equator to A at latitude θ decreases its distance from Earth's axis of rotation from R to $R\cos\theta$.

A6.2. The Coriolis Force

The Coriolis force can be explained by considering a ball thrown by a pitcher P from the center of a turntable toward a stationary catcher C off the turntable. The ball is thrown when a player H, on the perimeter of the turntable, happens to be on a straight line between P and C (fig. A6.2). By the time the ball reaches C, H has rotated to a new position. From the perspective of H, the ball started in his direction but then was deflected to the right of its motion. He can account for the motion of the ball by keeping in mind that he is rotating. Alternatively, if he prefers to ignore that he is on a turntable, he can assume that a Coriolis force deflects the ball. If we choose to ignore that we are on a rotating Earth, then, to explain the motion of parcels of air, we too must invoke a Coriolis force that (a) is proportional to the speed of the ball; (b) is to the right of its motion in the northern hemisphere, to the left in the southern hemisphere; (c) vanishes at the equator.

This explanation for the Coriolis force applies to parcels that move north-south. The same force comes into play when parcels move east-west, in circles around the axis of rotation. To appreciate why this is so, consider the forces that come into play when we stir the tea in a cup until it is rotating steadily. The surface of the tea starts as a horizontal plane (when the tea is motionless) and becomes a sloping parabola (when the tea is rotating). Because of this sloping surface, a fluid parcel experiences a pressure force directed toward the axis of rotation as shown in figure A6.3. This centripetal force enables the

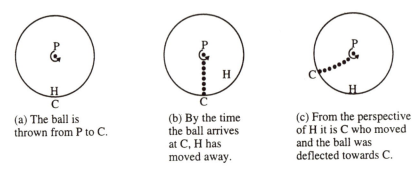

(a) The ball is thrown from P to C.

(b) By the time the ball arrives at C, H has moved away.

(c) From the perspective of H it is C who moved and the ball was deflected towards C.

Figure A6.2 A hitter H tries to intercept a ball thrown by pitcher P to catcher C. The pitcher and hitter are on a turntable, but the catcher is not.

parcel to continue moving in a circle (the way a ball can move in a circle provided it is at the end of a tense rope that continually pulls it toward the center of the circle). If we adopt a reference frame that rotates with the parcel, then the inward pressure force appears to be balanced by an outward centrifugal force that is proportional to U^2/R, where U is the speed of the parcel and R is the radius of the circle in which it moves. If the speed of rotation of the tea increases, so that the centrifugal force increases, then the slope of the tea surface becomes steeper so that the inward pressure force on the parcel also increases; the two forces stay balanced.

Let the rate of rotation be Ω so that a parcel at a distance R from the axis of rotation moves with speed ΩR. Consider an isolated parcel of tea that starts to rotate faster than the rest of the tea. The centrifugal force F is now increased because the parcel is moving faster. The pressure force is unchanged so that the parcel experiences a net outward force; it therefore starts to move outward, away from the axis of rotation. To an observer rotating with the tea, this can be explained by invoking a force, a Coriolis force, that deflects the parcel toward the right once the parcel starts to move. Similar arguments apply to motion on the surface of a rotating sphere.

In summary, the Coriolis force is present, irrespective of the direction of the motion. It is toward the right in the northern hemisphere, toward the left in the southern hemisphere, vanishes at the equator, and is proportional to the speed of a parcel. On a rotating Earth, we need to invoke this force to explain motion relative to us, whether it be the flow of air parcels that are deflected as they travel from zones of high pressure to zones of low pressure, or water parcels in the ocean that are deflected from the direction in which the wind drives them. (Icebergs in the northern hemisphere travel to the right of the direction of the prevailing wind.)

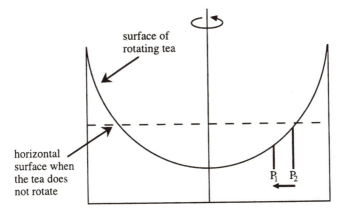

Figure A6.3 The pressure force from P_2 to P_1 permits a parcel to move in a circle. In a rotating frame of reference the pressure force is balanced by a centrifugal force.

A6.3. Shape of Earth

Earth is not a perfect sphere but has the shape of an oblate ellipsoid; it has an equatorial bulge because it rotates about an axis through the poles. A particle on the surface experiences a radially outward centrifugal force and, because of gravity, a much larger inward force that is directed, not toward the axis of rotation but toward Earth's center (fig. A6.4). The forces on a particle at the surface of a perfect sphere therefore have an equatorward component, which causes the surface to deform until it has a shape such that the direction of effective gravity is normal to the surface. (Effective gravity is the vector sum of the two forces acting on the parcel.)

A6.4. Gradient Winds

Consider cyclonic motion in a circle around a center of low pressure as shown in figure 6.8. In the absence of friction, a parcel of air is subject to two forces: a radially inward pressure force P_H and an outward Coriolis force that is proportional to the speed V of the parcel:

$$fV = P_H.$$

In this equation, which describes a geostrophic balance, f denotes the Coriolis parameter which, at a latitude θ, has the value $2\Omega\sin\theta$, where Ω is the angular velocity of Earth. Suppose that Earth were not rotating but that the parcel of air were still moving in a circle. In such

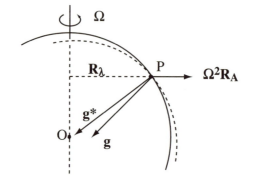

Figure A6.4 The centrifugal force on a particle P on Earth's surface is balanced by a gravitational force g^* toward Earth's center. The sum of these two vectors is the effective gravity g in the figure. The solid curve represents a sphere; the dashed line shows how the sphere is deformed into a geoid. From Wallace and Hobbs (1977).

a case, the pressure force would be balanced by a radially outward centrifugal force V^2/R. Hence, in response to a given pressure force, the speed of the parcel is determined, not by the previous equation, but by the equation

$$V^2/R + fV = P_H.$$

If R, the radius of the circle, is very small, then the Coriolis force is unimportant, and the circular motion associated with a given pressure force can be either clockwise or anticlockwise. Such is the case in a dust devil. An extratropical cyclone is usually so large—R has such a large value—that the centrifugal force is unimportant and the flow is in geostrophic balance. In the case of a hurricane, all three forces are important, and the motion is said to correspond to a gradient wind. This wind is weaker than the geostrophic winds in the case of cyclonic motion. The opposite is true when the flow is anticyclonic.

Exercises.

1. Explain the balance of forces on a geostationary satellite that is always above a fixed point on Earth's surface at the equator. Estimate the altitude of the satellite.

2. Identical twins, each weighing 200 kg, are on different trains on the equator. One train travels eastward at 20 m/sec, the other travels westward at the same speed. Estimate the resulting changes in the weights of the twins.

3. Explain why the winds in a small rotating dust devil, but not in a hurricane, can be either clockwise or anticlockwise.

Further Reading.   For a discussion of the interaction between climate and culture, see the article by Morgan (1989) and the book by Ross (1991). Peixoto and Oort (1992) provide a comprehensive discussion of the physics of Earth's climate. Wallace and Hobbs (1977) cover the topics of this chapter in detail.

APPENDIX 7

WEATHER, THE MUSIC OF OUR SPHERE

A7.1. Predicting the Weather

Keeping a record of the local weather over a period of several months is a useful exercise for several reasons. During the course of the measurements, consisting of at least twice-daily measurements over the spring semester—say, from January to May—the intensity of sunshine will increase greatly, causing seasonal global warming (assuming that the observer is in the northern extratropics). From the record, it will be evident that, because of superimposed short-term weather fluctuations, it is difficult to determine the onset of this warming. The natural variability of the atmosphere similarly complicates the documentation of gradual global warming associated with the increase in the atmospheric concentration of greenhouse gases.

Weather may appear chaotic if the only available information is local. Placed in the context of developments over a larger region, however, coherent patterns emerge. To learn about such patterns, plot the local data, plus those from other stations available on the World Wide Web, on a map such as the one in figure 7.6a. This is usually done by adopting the convention in figure A7.1.

The open circle in this figure indicates that the sky is clear; the arrow indicates the direction from which the wind is blowing, the northwest in this case; and the barbs or feathers indicate the wind speed, one for every 10 knots (1.95 knots = 1 m/sec). The numbers indicate a temperature of 56°F, dew point of 48°F, and a pressure of 1010 mb. In the case of the full circle, the sky is completely overcast. At that station, the wind is from the southwest, temperature is 66°F, dew point is 65°F, and pressure is 1000 mb. The two serrated lines indicate a warm and a cold front, respectively.

The next step is to draw isobars (lines of constant pressure) on the map. This exercise demonstrates how subjective judgments come into play because of the need to estimate the value of the pressure at points where no measurements are available. Values at nearby stations provide constraints that can be used in objective interpolation schemes. For example, it is usually assumed that the pressure at a point midway between two stations is the average of the pressure at

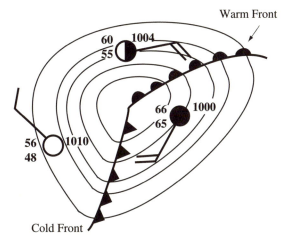

Figure A7.1   Typical symbols that appear on weather maps. Contours are isobars (lines of constant pressure) at intervals of 4 mb.

those two stations. Such objective schemes are based on assumptions that may not always be valid. For example, the surface pressure does not always vary smoothly and uniformly from one station to the next. The variation will not be uniform if there is a deep valley, or a small mountain between the stations, which explains why an individual familiar with a certain region can often prepare maps that are more accurate than those produced by a computer that follows an objective scheme.

A map can reveal that different air masses occupy different parts of the region under consideration. For example, a region with high surface temperature and high dew points is probably occupied by moist tropical air from the Gulf of Mexico. Cold arctic air, on the other hand, has a low temperature and an even lower dew point. The boundaries between such different air masses are usually designated by a line that represents a front at Earth's surface. Locating a front can be problematic in mountainous regions, where variations in surface temperatures can be associated with changes in elevation, and over the oceans, where variations in surface temperatures are small.

Further Reading.   See the texts by Ahrens (1994), Moran and Morgan (1991), Wallace and Hobbs (1977).

APPENDIX 8

THE OCEAN IN MOTION

A8.1. The Seasonal Thermocline

Whereas the atmosphere is heated from below so that convection readily occurs, the ocean is heated from above. Over most of the extraequatorial oceans, sea surface temperatures are at a maximum toward the end of summer. The decrease that starts in the autumn can create an unstable situation with warmer water below cold water at the surface. The resulting mixing because of convection is enhanced by the winds that effectively stir the ocean. By the end of the winter, the mixed layer near the surface extends to depths on the order of 100 m, whereafter heating at the surface creates a summer thermocline, as shown in figure A8.1. The ability of the ocean, more so than the land, to moderate the seasonal extremes of temperature depends more on the mixing of the upper ocean by the wind than it does on the different specific heats of land and water.

A8.2. The Perpetual Salt Fountain

The density, D, of the ocean depends primarily on temperature, T, and salinity S (which is usually measured in parts per thousand). The following equation is an approximate formula for this dependence:

$$D = D_0 + a(T - T_0) + b(S - S_0),$$

where $a = -.15 \text{ kg m}^{-3}\text{C}^{-1}$; $b = .78 \text{ kg m}^{-3}$ per part per thousand salinity; $D_0 = 1027 \text{ kg/m}^3$; $T_0 = 10°\text{C}$; $S_0 = 35$ psu. (At considerable depths, where pressures are large, the dependence of density on pressure must be taken into account.) It is possible to have a stable situation even though cold water is above warm water, provided the salinity varies in such a manner that density always increases with increasing depth. This equation describes the necessary relation between temperature and salinity.

In the subtropics, where evaporation exceeds precipitation, the temperature and salinity of the surface waters of the ocean often have values that exceed those of the water at depth. Consider what happens when, in such a region, water from a considerable depth is

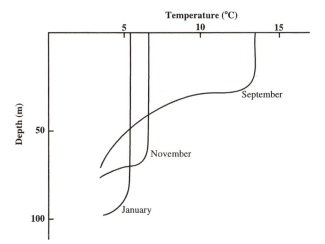

Figure A8.1   Temperature variations in the vertical, for different months, at a site in the northeastern Pacific Ocean. The thermocline is seen to be shallow and sharp at the end of the summer. During the winter, it deepens because the ocean loses heat to the atmosphere and because of stirring by the winds.

pumped to the surface through a copper pipe. As the water rises in the pipe, so does its temperature because copper conducts heat efficiently. Hence, the pipe will be filled with water that has the same temperature but that is less saline than water outside the pipe. It follows that, at the bottom of the pipe, pressure (which measures the weight of the vertical column of water) will be lower just inside than outside the pipe. In principal, this horizontal pressure force toward the bottom of the pipe can indefinitely sustain rising motion in the pipe, which becomes a perpetual fountain that depends on the different salinities of the water at the surface and at depth.

Further Reading.   For a more detailed discussion of many of the topics introduced in this chapter, see the volume of articles edited by Warren and Wunsch (1981) and the texts by Gill (1982) and Philander (1990).

APPENDIX 9

EL NIÑO, LA NIÑA,

AND THE SOUTHERN OSCILLATION

The Southern Oscillation is but one phenomenon that depends on interactions between the ocean and atmosphere. Others include the curious climatic asymmetries relative to the equator in the eastern Pacific and Atlantic and the peculiar seasonal cycle of those regions.

Even though the time-averaged sunshine incident on Earth is perfectly symmetrical about the equator, Earth's climate is not: in the eastern equatorial Pacific and Atlantic Oceans, the doldrums—where rainfall is at a maximum—are well to the north of the equator. Thus, the neighborhood of Panama City at 10°N is verdurous, but the coastal zone of Peru, at the same distance from the equator in the southern hemisphere, is a barren desert. This asymmetry, which is essentially absent from the Indian Ocean and the western tropical Pacific and Atlantic Oceans, must be a consequence of the geometries of the continents. Two factors come into play. One is the global distribution of land masses, the other is the geometry of coastlines in certain regions. The first factor causes monsoonal winds to be dominant over the Indian Ocean and affects the trade winds over the tropical Atlantic and Pacific Oceans in a more subtle manner. The trades would be present on a water-covered globe; in the presence of continents, the temperature contrasts between land and ocean intensify those winds. In the oceans, those winds create east-west sea surface temperature gradients that in turn cause a further intensification of the winds. In other words, the time-averaged winds influence and are also influenced by the sea surface temperature patterns. The implied interactions between the ocean and atmosphere would probably be symmetrical about the equator, were it not for the details of the geometries of the western coasts of Africa and the Americas. Because it attains temperatures higher than those of the adjacent ocean, the bulge of west Africa north of the equator favors winds with a southerly component over the eastern Atlantic. Those winds induce upwelling along the southwestern African coast, thus causing the climatic asymmetry. In the eastern tropical Pacific, the reason for the climate asymmetries relative to the equator is different. The prevailing winds are anticyclonic around the offshore high-pressure zones. Because of the slope of the western coast of the Americas relative to meridians, the winds are parallel to the coast south of the equator, less so north of the equator. Hence, upwelling and low sea surface temperatures are most pronounced just south of the equator.

Even though the sun "crosses" the equator twice a year, sea surface temperatures on the equator in the eastern equatorial Pacific and Atlantic have maxima once a year, in March and April. The seasonal cycle of the southern hemisphere in effect extends across the equator into the northern hemisphere. The main reason for this phenomenon is the asymmetry of the time-averaged climate mentioned above. In the absence of that asymmetry, the winds at the equator would blow toward the summer hemisphere: northward during the northern summer, southward during the southern summer. Because the time-averaged winds blow northward at the equator, toward the doldrums, the superimposed seasonal cycle causes the winds to vary seasonally from intense (in late northern summers) to weak (in late southern summers), without changing direction. Hence, evaporation from the ocean is strong and sea surface temperatures are low during northern summers, but evaporation is weak and temperatures are high during southern summers.

Further Reading.   The National Research Council Report prepared by its TOGA Panel is an excellent and readable summary of recent developments in the field of ocean-atmosphere interactions. The fine review article by Neelin et al. (1994) is technical, as is the book by Philander (1990).

APPENDIX 10

THE PARADOX OF THE FAINT SUN

BUT WARM EARTH

A10.1. Weathering

Weathering depends on extremely complex chemical reactions; the following is a simplified representation that captures the essential processes. The removal of carbon dioxide from the atmosphere involves a reaction of that gas with rainwater to form carbonic acid.

$$CO_2 + H_2O \rightarrow H_2CO_3$$

The acidic rain dissolves both silicate and carbonate rocks, such as limestone, into calcium and bicarbonate ions:

$$CO_2 + H_2O + CaCO_3 \rightarrow Ca^{++} + 2HCO^-_3$$
$$2CO_2 + H_2O + CaSiO_3 \rightarrow Ca^{++} + 2HCO^-_3 + SiO_2.$$

Streams and rivers carry these ions to the oceans where marine organisms construct skeletons of carbonates, especially calcium carbonate:

$$Ca^{++} + CO^{--}_3 \rightarrow CaCO_3.$$

When these organisms die they fall to the ocean floor and their remains are incorporated into sediment that, in due course, is recycled through the solid Earth. Under high temperatures and pressure, carbonates and silica react to form silicate minerals and carbon dioxide:

$$CaCO_3 + SiO_2 \rightarrow CaSiO_3 + CO_2,$$

where $CaSiO_3$ represents a generic silicate mineral. The carbon dioxide produced in this manner can be returned to the atmosphere by volcanoes.

Further Reading. The article by Kasting et al. (1988) describes the evolution of the terrestrial planets.

A10.2. Properties of the Planets

TABLE A10.1

Planet	Average Distance from Sun (A.U.)[a]	Orbital Eccentricity	Inclination of Equator to Orbit (degrees)	Length of Year (days)	Period of Rotation (days)	Mass (Earth = 1)	Mean Radius (km)	Gravitational Attraction cm sec^{-2}
Mercury	0.39	.206	0	8	58.7	0.054	2439	376
Venus	0.72	.007	<3	225	−243[c]	0.81	6049	888
Earth	1.00	.017	23.5	365	1.00	1.00[b]	6371	981
Mars	1.52	.093	25.2	687	1.03	0.11	3390	373
Jupiter	5.2	.048	3.1	4330	0.41	314	69,500	2620
Saturn	9.5	.056	26.8	10,800	0.43	94	58,100	1120
Uranus	19.2	.047	98.0	30,700	−0.89[c]	14.4	24,500	975
Neptune	30.1	.009	28.8	60,200	0.53	17.0	24,600	1134

[a]Astronomic Unit = the mean distance of the Earth from the Sun = 149.6 × 10⁶ km.
[b]The mass of the Earth = 5.97 × 10²⁷ g.
[c]Venus and Uranus rotate in the opposite sense to the other planets.

APPENDIX 11

WHY SUMMER IS WARMER THAN WINTER:

EARTH'S SENSITIVITY TO PERTURBATIONS

The ice ages provide information about the sensitivity of Earth to perturbations in the form of changes in the distribution of sunlight. The results are therefore not directly applicable to the climate changes that are likely because of a different perturbation, an increase in the atmospheric concentration of greenhouse gases. It is particularly important to note that variations in the tilt of Earth's axis, and in the precession of its axis, cause no net change in the sunlight incident on Earth, only in the global and seasonal distribution of the sunlight. (A change in the tilt causes some latitude bands to receive more sunlight and others to receive less. The precession causes sunlight to be more intense during certain months, less intense during other months.) Changes in the eccentricity of Earth's orbit do cause a net change in the incident sunlight. At this time, it is not known why Earth's response to this extremely modest perturbation is particularly large.

Exercises.

1. In 1714, the British Parliament enacted the Longitude Act, which promised a prize of £20,000 (millions of dollars today) for a solution to the problem of determining longitude at sea. (The problem contributed to numerous ship wrecks, because a captain who does not know how far to the east or west his ship has traveled is uncertain of the distance to land.) The prize finally went to John Harrison, an unschooled woodworker who developed a pendulum-free clock that could keep accurate time at sea. Explain why the invention of a clock provides a solution to the problem of determining longitude. How would you determine latitude at sea?

2. Explain the terms Tropic of Cancer, Tropic of Capricorn, the equinoxes, the Solstices, and the Arctic Circle.

3. Discuss briefly the climate of a planet that is identical to Earth except that its axis of rotation is vertical and inclined at an angle of 45°.

Further Reading. The engaging book by Imbrie and Imbrie (1979) describes scientific explorations of the ice ages. For a more technical account of Earth's past climates, see Crowley and North (1991).

APPENDIX 12

THE OZONE HOLE, A CAUTIONARY TALE

The concentration of ozone in the stratosphere cannot be explained strictly in terms of creation (from oxygen molecules and atoms) and destruction by energetic photons. In the 1970s, Crutzen identified important additional reactions. They involve the decay of nitrous oxide that produces NO and NO_2, which react catalytically (without themselves being consumed) with ozone:

$$NO + O_3 \rightarrow NO_2 + O_2$$
$$O_3 + photon \rightarrow O_2 + O$$
$$NO_2 + O \rightarrow NO + O_2.$$

The net result is the conversion of two ozone molecules into three oxygen molecules:

$$2O_3 \rightarrow 3O_2.$$

Further Reading. Benedick (1991) describes negotiations that led to the Montreal Protocol. The texts by Turco (1997) and Graedel and Crutzen (1993) include comprehensive discussions of ozone chemistry.

APPENDIX 13

GLOBAL WARMING, RISKY BUSINESS

Further Reading. The reports of the International Panel on Climate Change, edited by Houghton et al. (1990, 1995) cover all aspects of global warming. For a discussion of the carbon cycle, see the articles by Sarmiento (1993) and Post et al. (1990).

GLOSSARY

Absolute zero — The temperature at which, theoretically, there is no molecular motion: $-273°C$, $-460°F$, or $O°K$.

Adiabatic process — A process that involves no transfer of heat. An air parcel that is moved adiabatically exchanges no heat with its surroundings. Should this movement result in compression of the parcel, then its temperature will rise; if the parcel expands, then the result is cooling.

Aerosols — Tiny suspended solid particles (dust, smoke, etc.) or liquid droplets in the atmosphere.

Albedo — The brightness of a surface, which determines what fraction of any incident light it reflects (and hence does not absorb). Bright surfaces such as snow, have a high albedo; dark surfaces have a low albedo.

Antarctic Circumpolar Current — An eastward current that circles the globe at high southern latitudes; the only current that connects Earth's three major ocean basins.

Anticyclone — An area of high pressure around which the wind blows clockwise in the northern hemisphere and counterclockwise in the southern hemisphere.

Blackbody — A hypothetical object that absorbs all the radiation that is incident on it.

Carbon dioxide (CO_2) — A colorless, odorless gas that effectively absorbs infrared radiation and thus is a powerful greenhouse gas. Its atmospheric concentration, about 0.035% (355 ppm) near sea level, is rising rapidly because of our industrial and agricultural activities. Solid carbon dioxide is called dry ice.

Centripetal force — The radial force required to keep an object moving in a circular path. It is directed toward the center of that curved path.

Chinook wind — A warm, dry wind on the eastern side of the Rocky Mountains. In the Alps, this wind is called a *foehn*.

Chlorofluorocarbons (CFCs) — Chemical compounds composed of carbon, hydrogen, chlorine, and fluorine; once widely used as aerosol propellants but banned after they were found to contribute to ozone depletion in the stratosphere.

Cirrus — A high, thin, featherlike cloud composed of ice crystals.

Condensation — The process by which the gas water vapor becomes a liquid.

Conduction — The transfer of heat by molecular activity; this process requires physical contact for the flow of heat from a region of high temperature to a region of low temperature.

Convection — Motion in a fluid that results in the transport and mixing of the fluid's properties. Convection occurs when the density of a fluid increases with height. This usually happens because of heating from below. In the ocean, evaporation or the formation of icebergs can result in convection by increasing the salinity and hence density of the surface waters.

Coriolis force — The apparent force that deflects a freely moving object in a rotating system. On Earth, the deflection of parcels of air in the atmosphere, and of parcels of water in the ocean, is to the right in the northern hemisphere and to the left in the southern hemisphere.

Cyclone — An area of low pressure around which the winds blow counterclockwise in the northern hemisphere and clockwise in the southern hemisphere.

Dew — Water that has condensed onto objects with temperatures below the dew point of the surrounding air.

Dew point temperature — The temperature to which air must be cooled (at constant pressure and constant water vapor content) for saturation to occur.

Doldrums — The overcast, rainy region of light winds, near the equator (see Intertropical Convergence Zone).

Easterlies — Winds from the east, the trade winds of the tropics, for example.

Eccentricity (of Earth's orbit) — The degree to which Earth's elliptical orbit departs from a circle.

El Niño — Interannual oceanic warming of the eastern tropical Pacific off the coasts of Peru and Ecuador. It is one phase of the Southern Oscillation. The other phase is La Niña.

Equatorial Undercurrent — An intense, subsurface eastward oceanic jet that flows precisely along the equator in the Pacific and Atlantic Oceans. In the Indian Ocean, it appears briefly in March and April.

Equinox — The time, March 21 and September 21, when the sun is directly over the equator.

Euphotic zone — The uppermost layer of a body of water that receives enough sunlight to support plant life.

Evaporation — The process by which a liquid changes into a gas.

Feedback mechanism — A process whereby the response to an initial change either reinforces the change (positive feedback) or weakens it (negative feedback).

Fog — A cloud with its base at Earth's surface.

Front — The transition zone between two distinct air masses. Along a cold front, a mass of cold air is advancing and replacing a mass of warm air. A warm front moves in such a way that warm air replaces cold air.

Geostationary satellite — A satellite that remains over a fixed placed above the equator because it orbits Earth at the same rate that Earth rotates.

Greenhouse effect — The warming of a planetary surface that results when the atmosphere absorbs and reemits infrared radiation. The principal greenhouse gases in Earth's atmosphere are water vapor and carbon dioxide.

Gulf Stream — The warm current that flows along the eastern coast of the United States before it veers offshore at Cape Hatteras and continues northeastward toward Europe.

Hadley cell — A thermal circulation, proposed by George Hadley, that involves moist air that rises in low latitudes where surface temperatures are high, and dry air that subsides in the neighborhood of 30° latitude.

Humidity — *Absolute humidity* measures the mass of water vapor in a given volume of air. *Relative humidity* measures the ratio of the amount of water vapor actually in the air compared to the amount of water vapor the air can hold at that particular temperature and pressure. This is the same as the ratio of the air's actual vapor pressure to its saturation vapor pressure.

Infrared radiation — Electromagnetic radiation at wavelengths between approximately 0.7 and 1000 micrometers, which is longer than the wavelengths of visible light.

Insolation — The solar radiation that reaches Earth.

Intertropical Convergence Zone (ITCZ) — The region of rising air, cloudiness and heavy rainfall onto which the northeast trade winds of the northern hemisphere, and the southeast trade winds of the southern hemisphere, converge.

Inversion — An increase in air temperature with height.

Isobar — A line of constant pressure.

Jet Streams — The bands of intense westerly winds in the midlatitudes of each hemisphere.

Kuroshio current — A swift, warm poleward current that flows along the eastern shores of Japan before it penetrates the Pacific.

La Niña — Interannual cooling of the eastern tropical Pacific. This is one phase of the Southern Oscillation; the complementary phase is El Niño.

Lapse rate — The rate at which atmospheric temperatures decrease with height near Earth's surface.

Latent heat — The heat that is either released or absorbed by a unit mass of a substance when it undergoes a change of state, for example, during evaporation, condensation, or sublimation.

Milankovitch theory — A theory proposed by Milutin Milankovitch in the 1930s to explain the recurrent ice ages as a consequence of periodic changes in Earth's orbital parameters.

Millibar (mb) — A unit for expressing atmospheric pressure. Sea-level pressure is normally close to 1013 mb.

Obliquity (of Earth's axis) — The tilt of Earth's axis relative to the plane of its orbit.

Ozone (O_3) — An almost colorless gaseous form of oxygen that is abundant in the stratosphere.

Photon — The "particle" of electromagnetic energy; it has no electrical charge.

Phytoplankton — Tiny marine plants that drift near the surface of the ocean.

Plate tectonics — The theory that Earth's surface is divided into a number of plates that move relative to one another. Also referred to as continental drift.

Precession (of Earth's axis of rotation) — The wobble of Earth's axis of rotation that traces out the path of a cone.

Radiant energy (radiation) — Energy propagated in the form of electromagnetic waves that can also be viewed as a stream of discrete photons.

Refraction — The bending of light as it passes from one medium to another, air into water, for example.

Solar Wind — The stream of charged particles, including protons and electrons, that the sun emits, and that travels at speeds of 400 to 500 km/sec.

Solstice — The time when the sun is directly overhead at either the tropic of Cancer (latitude 23.5°N on June 22 approximately) or the tropic of Capricorn (latitude 23.5°S on December 22). June 22 is the summer solstice for the northern hemisphere, the winter solstice for the southern hemisphere.

Southern Oscillation — The irregular interannual climate fluctuation between El Niño and La Niña states. It involves a see-saw in the surface pressure across the tropical Pacific Ocean.

Stratosphere — The layer of the atmosphere, between 10 km and 50 km approximately, in which temperatures increase with height. It is above the troposphere and below the mesosphere.

Sublimation — The process whereby ice changes directly into water vapor without melting.

Temperature — The degree of hotness or coldness of a substance as measured by a thermometer. It is also a measure of the average speed or kinetic energy of the atoms and molecules in a substance.

Thermocline — The subsurface region of large temperature gradients that separates the warm surface waters of the ocean from the cold water at depth.

Thermohaline circulation — This circulation, which maintains low temperatures in the deep ocean, involves the sinking of high-density, cold, saline surface water around Antarctica and in high latitudes of the northern Atlantic.

Trade winds — The easterly winds that prevail over most of the tropics.

Tropopause — The boundary between the troposphere and the stratosphere.

Troposphere — The layer of the atmosphere that extends from Earth's surface to the tropopause.

Typhoon — A hurricane over the western Pacific Ocean.

Ultraviolet radiation — Electromagnetic radiation with wavelengths longer than x rays but shorter than visible light.

Upwelling — Upward oceanic motion that brings cold nutrient-rich water to the surface.

Vapor pressure — The pressure exerted by the water vapor molecules in a given volume of air.

Visible radiation (light) — Radiation with a wavelength between approximately 0.4 and 0.7 micrometers.

Walker circulation — An atmospheric cell that involves the rising of moist air over the warm western tropical Pacific, its subsidence as dry air over the cold eastern tropical Pacific, and its return to the west in the trade winds.

Water vapor — Water in a vapor (or gaseous) form.

Wavelength — The distance between successive crests, troughs, or identical parts of a wave.

Westerlies — The winds from the west that prevail at Earth's surface in midlatitudes. The roaring forties and screaming fifties of the southern hemisphere refer to these winds.

REFERENCES

Ahrens, C. D. 1994. *Meteorology Today: An Introduction to Weather, Climate, and the Environment.* 5th ed. Minneapolis/St. Paul: West Publishing.

Barnola, S. M., D. Raynaud, Y. S. Korotkevich, and C. Lorius. 1987. *Nature* 329:408.

Benedick, R. E. 1991. *Ozone Diplomacy: New Directions in Safeguarding the Planet.* Cambridge, Mass.: Harvard University Press.

Bentley, W. A., and W. J. Humphreys. 1931. *Snow Crystals.* New York: Dover.

Broecker, W. S. 1985. *How to Build a Habitable Planet.* New York: Eldigio Press.

Brown, G., Jr. 1993. *Chemical and Engineering News* 71(22):9–11.

Buzyna, G., R. C. Pfeffer, and R. Kung. 1984. *Journal of Fluid Mechanics* 145:377–403.

Colinvaux, P. 1978. *Why Big Fierce Animals Are Rare: An Ecologist's Perspective.* Princeton, N.J.: Princeton University Press.

Crowley, T. J., and G. R. North. 1991. *Paleoclimatology.* New York: Oxford University Press.

Emanuel, K. A. 1994. *Atmospheric Convection.* New York: Oxford University Press.

Friedi, H., H. Lotscher, H. Oeschger, U. Siegenthaler, and B. Stauffer. 1986. *Nature* 324:237–238.

Gill, A. E. 1982. *Atmosphere-Ocean Dynamics.* New York: Academic Press.

Glaisher, J. 1871. *Travels in the Air.* London: Richard Bentley and Son.

Graedel, T. E., and P. J. Crutzen. 1993. *Atmospheric Change, an Earth System Perspective.* New York: W. H. Freeman.

Gribben, J. 1984. *In Search of Schrodinger's Cat (Quantum Physics and Reality).* New York: Bantam Books.

Hardin, G. 1968. *Science* 162:1243–1248.

Hays, J. D., J. Imbrie, and N. J. Shackleton. 1976. *Science* 194:1121–1132.

Holmes, R. 1994. *Science* 264:1252–1253.

Houghton, J. T., G. J. Jenkins, and J. J. Ephraums, eds. 1990. *Climate Change: The IPCC Scientific Assessment.* New York: Cambridge University Press.

———, eds. 1995. *Climate Change.* New York: Cambridge University Press.

Imbrie, J., and K. P. Imbrie. 1979. *The Ice Ages: Solving the Mystery.* Cambridge, Mass.: Harvard University Press.

Intergovernmental Panel on Climate Change (IPCC). 1992. *Climate Change 1992: The Supplementary Report to the IPCC Scientific Assessment.* New York: Cambridge University Press.

Kasting, J. F., O. B. Toon, and J. B. Pollack. 1988. *Scientific American* 260:90–97.

Kennedy, P. 1993. *Preparing for the Twenty-First Century.* New York: Random House.

Kiehl, J. T., and K. E. Trenberth. 1997. *Bulletin of the American Meteorological Society* 78:197–208.

Kirchner, J. W. 1989. *Reviews of Geophysics* 27:223–235.

Knauss, J. A. 1997. *Introduction to Physical Oceanography.* 2d ed. Englewood Cliffs, N.J.: Prentice-Hall.

Lorenz, E. N. 1993. *The Essence of Chaos.* Seattle: University of Washington Press.

Lorius, C., J. Jouzel, D. Raynaud, J. Hansen, and H. Le Treut. 1990. *Nature* 349:139.

Lovelock, J. E. 1988. *The Ages of Gaia: A Biography of Our Living Earth.* New York: W. W. Norton.

Malthus, T. R. 1798. *An Essay on the Principle of Population As It Affects the Future Improvement of Society.* London: Ward, Lock.

Marotzky, J., and J. Willebrand. 1991. *Journal of Physical Oceanography* 21:1372–1385.

Moran, K. M., and M. D. Morgan. 1991. *Meteorology, the Atmosphere, and the Science of Weather.* 3d ed. New York: Macmillan.

Morgan, D. 1989. *Oceanus* 32(2):47–49.

National Academy of Sciences, Committee on Science Engineering, and Public Policy, Synthesis Panel. 1992. *Policy Implications of Greenhouse Warming.* Washington, D.C.: National Academy Press.

National Research Council Advisory Panel for the Tropical Oceans and Global Atmosphere (TOGA) Program. 1996. *Panel Report.* E. Sarachik, chairman. Washington, D.C.: National Academy Press.

Neelin, J. D. 1990. *Journal of Atmospheric Sciences* 47(5):674–693.

Neelin, J. D., F-F. Jin, and M. Latif. 1994. *Annual Review of Fluid Mechanics* 26:617–659.

Neftel, A., E. Moore, H. Oeschger, and B. Stauffer. 1985. *Nature* 315:45–47.

Peixoto, J., and A. H. Oort. 1992. *Physics of Climate.* Forward by Edward M. Lorenz. New York: American Institute of Physics.

Philander, S. G. H. 1990. *El Niño, La Niña, and the Southern Oscillation.* New York: Academic Press.

Post W. M., T-H Peng, W. R. Emanuel, A. W. King, V. H. Dale, and D. L. De Angelis. 1990. *American Scientist* 78:310–326.

Rasmussen, E. M., and J. M. Wallace. 1983. *Science* 222:1195–1202.

Ross, A. 1991. *Strange Weather.* New York: Verso.

Rowland, F. S. 1990. *Ambio* 19(6–7):281–292.

Sarmiento, J. L. 1993. *Chemical and Engineering News* 71:30–43.

Siegenthaler, U., H. Friedli, H. Loetscher, E. Moor, A. Neftel, H. Oeschger, and B. Stauffer. 1988. *Annals of Glaciology* 10:1–6.

Stone, R. 1994. *Science* 264:1252–1253.

Trenberth, K. E. 1992. *Climate System Modeling.* New York: Cambridge University Press.

Turco, R. P. 1997. *Earth under Siege: From Air Pollution to Global Change.* New York: Oxford University Press.

Turekian, K. K. 1996. *Global Environmental Change: Past, Present, and Future.* Englewood Cliffs, N.J.: Prentice-Hall.

Wallace, J. M., and P. V. Hobbs. 1977. *Atmospheric Science.* New York: Academic Press.

Warren, B. A., and C. Wunsch, eds. 1981. *Evolution of Physical Oceanography.* Cambridge, Mass.: MIT Press.

INDEX